George Gabriel Stokes

On Light

First Course, on the Nature of Light, Delivered at Aberdeen in November, 1883

George Gabriel Stokes

On Light

First Course, on the Nature of Light, Delivered at Aberdeen in November, 1883

ISBN/EAN: 9783337250041

Printed in Europe, USA, Canada, Australia, Japan

Cover: Foto ©berggeist007 / pixelio.de

More available books at **www.hansebooks.com**

BURNETT LECTURES.

ON LIGHT.

BURNETT LECTURES.

ON LIGHT.

First Course,

ON THE NATURE OF LIGHT,

DELIVERED AT ABERDEEN IN NOVEMBER, 1883,

BY

GEORGE GABRIEL STOKES, M.A., F.R.S. &c.

FELLOW OF PEMBROKE COLLEGE, AND LUCASIAN PROFESSOR OF
MATHEMATICS IN THE UNIVERSITY OF CAMBRIDGE.

London:
MACMILLAN AND CO.
1884

Cambridge:

PRINTED BY C. J. CLAY, M.A. & SON,

AT THE UNIVERSITY PRESS.

PREFACE.

THE lectures, of which the first course is contained in the present volume, originated in a new direction given to an old endowment. Mr John Burnett, merchant in Aberdeen, who died in 1784, bequeathed the greater part of his property to various charitable and pious objects. Among others, a portion of the property was vested in trustees for establishing prizes for the best and next best essay on the following subject :—

"That there is a Being, all-powerful, wise, and good, by whom everything exists; and particularly to obviate difficulties regarding the wisdom and goodness of the Deity; and this, in the first place, from considerations independent of written revelation,

S.   *b* and in the second place, from the revelation of the Lord Jesus; and, from the whole, to point out the inferences most necessary for, and useful to mankind."

These essays were to be competed for at intervals of 40 years, and awards have been made on two occasions since the original foundation.

But it was thought that the production of essays at such long intervals did not form a satisfactory mode of utilizing the bequest of the founder; and in 1881 a new direction was given to the foundation by an order of the Secretary of State for the Home Department, in accordance with the provisions of the Endowed Institutions (Scotland) Act of 1878. By this order it was provided that a Lecturer should be appointed at intervals of 5 years, to hold office for 3 years, the subject of the lectures being determined by the following regulation :—

"The Trustees and assessors may prescribe as the subject of the course of lectures that specified in the codicil to the will of the testator, viz. [as above]. Or otherwise, the trustees may prescribe, as the subject of any course of lectures, recent researches (as at the date of the appointment) into any of the following branches of knowledge, viz :—

1. History, including the illustrations of the forms and effects of theistic doctrines among the older nations of the world.

2. Archaeology.

3. Physical science.

4. Natural science.

And they shall instruct the lecturer to have regard, in treating of the special subject prescribed, to the illustration afforded by it of the theme proposed by the testator, and that under such conditions or qualifications as they may prescribe."

The Trustees selected physical science as the subject of the first lectures under the new system, and they did me the honour of appointing me the first Burnett Lecturer. It has been arranged that four lectures are to be given each year of office, and that the subject of the complete course shall be Light, the subject being treated under three divisions, to which the lectures in each year are to be respectively devoted. The first

division is sufficiently explained by the title of
the present course ; the subjects of the second
and third divisions will be found at the end of
the fourth lecture.

I am sensible of the difficulty of the task I
have attempted in this first year's course,
namely, while avoiding all mathematical details,
to give the audience some fair idea of the evidence on which we accept the views respecting
the nature of light which are at present held, I
may almost say universally, in the scientific
world. I assume on the part of the reader a
knowledge of the rectilinear propagation of light
in the same medium, of the laws of reflection and
refraction, of the compound character and of the
decomposition of white light ; but I have not
assumed that he is acquainted with the phenomena of interference, or diffraction, or double
refraction, or polarization, though some acquaintance with these subjects will make the lectures
much more easy to follow. For the sake of
those persons to whom these things may be

nearly or entirely new, I have been obliged to enter into details which I cannot help fearing may be wearisome to those who have already a pretty good acquaintance with the subject; though I hope that even they may not find the weighing of the evidence wholly uninteresting or unprofitable.

G. G. STOKES.

CAMBRIDGE,
February 5, 1884.

CONTENTS.

LECTURE I.

PAGE

Bearing of the more obvious properties of Light on our view of its nature—Theories of emission and of undulations—Colours of thin plates—Newton's attempt to explain them on the theory of emission—Insufficiency of the explanation   1

LECTURE II.

Interference—Explanation of the colours of thin plates afforded by the theory of undulations—Diffraction   32

LECTURE III.

Closer examination of the fundamental suppositions of the Theory of Undulations—Survey of the conclusions arrived at by a study of the phenomena of common light—Elementary facts of double refraction and polarization   74

LECTURE IV.

Phenomena presented on interposing a crystalline block or thin plate in the path of Polarized Light which is subsequently analyzed—Laws on Interference of Polarized Light—Theory of Transverse Vibrations—Conclusion   105

LECTURES ON LIGHT.

FIRST COURSE.

ON THE NATURE OF LIGHT.

LECTURE I.

Bearing of the more obvious properties of Light on our view of its nature—Theories of emission and of undulations—Colours of thin plates—Newton's attempt to explain them on the theory of emission—Insufficiency of the explanation.

AMONG all our senses, there is none more wonderful than that of sight. It confers upon us, as Sir John Herschel has remarked, to a considerable extent the character of ubiquity. It is accordingly a matter of extreme interest to find out what we can as to the mode in which this end so important to our well-being is brought about; whether it be by investigating the properties of any agent external to ourselves which may be concerned in its accomplishment, or by seeking to penetrate some little way into that

S.   I

mysterious chain of sequence which connects the external agent with the sensation conveyed to our minds. It is true indeed that there is no prospect of our being able to bridge over the gulf which separates mind from matter; yet there are many things to indicate that that mysterious organ which we possess, the brain, has some intimate connexion with the operations of the mind, and we can do something towards tracing a connexion between the part of our bodies directly affected by the external agent and the brain; a connexion of such a kind as to leave no doubt that it forms the means whereby the immediate action of the external agent is ultimately perceived by the mind; though how the final conveyance takes place is a mystery we are not likely to fathom. But this does not prevent us from being able to trace some links in the chain of connexion, nor from recognizing the evidences of design which that portion which we can in a measure follow is calculated to impress upon us.

I have spoken of an external agent even though it was more especially the sense of sight that I had in view. We are so accustomed to the contemplation of an objective *something*, which we call Light, as the external agent by the action of which vision is in some way brought about, that we have a difficulty in conceiving how anyone could think otherwise. We see the objects in a room, but it is matter of the commonest observation that though the objects may be there and our eyes may be there, we see them not if it be night, or if the room be closely shut up, until a lighted lamp or candle or something of the kind is introduced. We recognize the flame as the seat of some influence, to which we give the name of light, which is essential to vision. Yet obvious as this proposition appears to us, it is remarkable that it was not always so. At least one writer of acute intellect in ancient times held that it was something emanating from the eye, not something from without entering into it, that enabled us to perceive distant objects. This shows by what a slow and gradual progress our knowledge of physical science is built up. We are accustomed to vaunt of our knowledge in this 19th century; yet it may be that generations hence the scientific men of the day will wonder how we could have failed to perceive things which to them will appear quite obvious.

A self-luminous body then is, as such, a source of an influence which can be exerted at a distance. In this respect it by no means stands alone; two bodies at a distance from one another may for example influence each other through the attraction of gravitation, and other modes of influence might also be mentioned, but I will confine myself to gravitation.

I—2

Now between these two modes of influence there are great and striking differences. One of the most salient is this, that in the case of gravitation the influence is exerted independently of the interposition of matter of any kind, whereas in the case of light the influence is capable of being arrested by an extremely thin screen of matter of a suitable kind; for example, silver foil, or a film of Indian ink spread on glass.

The consideration of a screen leads us naturally to another fundamental and very obvious property of light. Suppose the screen pierced by one or more rather small apertures; how will the influence beyond the screen be distributed? We find that it is perceived only within the projections of the aperture or apertures made by straight lines drawn from the luminous body, which for simplicity I here suppose to be a point.

This indeed is not rigorously true, for about the boundary of the projection of the aperture, that is, near the projection of the edge of the aperture, there is a gradual, not sudden, passage from light within to darkness outside, accompanied by fluctuations of greater and less luminosity which I cannot now enter into particularly; and when the aperture is extremely small, the spreading out of the light which passes through it is by no means very small compared with the breadth of the projection of the aperture itself. Nevertheless in ordinary cases the spreading out of the light is so small that we may disregard it, and say that light proceeds from a luminous point in all directions in straight lines until it is stopped by some obstacle. It will be understood that I am here speaking of light only as it passes in free space, or which comes to nearly the same thing, in air, and not from one medium into another.

When I speak of light proceeding *from* a luminous point, all I wish to express is that we recognize the luminous point as the seat or origin of a certain influence which is exerted, though with an intensity which diminishes with increasing distance, at all points from which a straight line can be drawn without interruption to the luminous point ; I do not wish to imply the idea of motion, or propagation of any kind. The feature therefore that we are at present considering is common to light and gravitation.

A very important question now arises, Do these influences take time to travel, or does the influence exerted at any moment of time depend solely on the state and position of the two bodies, the influencing and the influenced, at that moment ? The idea which we may be led to form as to the nature of light must depend most materially on the answer we have to give to this question as applied to light.

That light is propagated in time, cannot be inferred from ordinary observation. Thus when a landscape is illuminated at night by a flash of lightning or an explosion of gunpowder, the flash and the objects illuminated by it are seen, so far as our senses can decide, simultaneously, though in the latter case the light has to travel from the flash to the object illuminated, and from that to the eye, instead of coming straight from the flash. As far as ordinary observation goes, then, the question whether any time is occupied in the transmission is left an open one, and we can only say that *if* time is required for transmission, the rate of travelling must be enormously great.

But in the solar system we have distances to deal with compared with which the dimensions of the earth itself on which we dwell, let alone those of a landscape, sink into insignificance. It is conceivable that in travelling over those vast distances, if it does travel at all, light might occupy lengths of time which would not be insensibly small, and which possibly might be put in evidence by some celestial phenomenon. Thus if the light of the sun were emitted by flashes, the sun would take the place of the thunder-cloud in our supposed observation, and the planets that of the objects in the landscape. Though there are changes going on in the sun, as we now know, there are none of such magnitude and suddenness as to be available for such an observation. Nevertheless it was by celestial observations, of a somewhat different kind, that the finite velocity of light was first revealed.

This was done as long ago as in 1676 by Roemer, who showed that an inequality in the times of occurrence of the eclipses of Jupiter's satellites, which he had observed, was simply explicable on the supposition that light is propagated with a finite velocity. The mean motion of a satellite round the planet being accurately known, from observations extending over a sufficient time, the times of successive eclipses can be calculated. It was found that the eclipses happened earlier or later than the calculated times, supposing the epoch so chosen as to make the accelerations and retardations balance on an average, according as the earth and Jupiter were on the same side of the sun or on different sides, and the differences in the apparent errors in the times of occurrence of the eclipses at different times of the year were proportional to the differences of the distances of Jupiter from the earth at those times. The greatest difference in that distance is evidently equal to the diameter of the earth's orbit (supposing for simplicity the orbits of the earth and Jupiter circular and in the same plane), and to this corresponds the greatest difference in the apparent errors of the times of occurrence, which amounts to about 16 minutes and a quarter. We learn therefore that if this be the true reason of the observed inequality, light takes 8 minutes and a few seconds to travel over the distance from the earth to the sun. To express the velocity in miles per second, we require to know the dimensions of the earth's orbit, or what comes to the same the solar parallax, which, being the ratio of the radii of the earth and the earth's orbit, serves to express the latter in miles since the former is accurately known. According to the value of the sun's parallax which was accepted till recent years, the velocity of light so determined came out about 192,000 miles per second. More recent determinations have reduced this to about 186,000 miles per second.

For nearly 50 years after Roemer's discovery, no other phenomenon was known which indicated that the propagation of light was other than instantaneous; but then Bradley made the very remarkable discovery of the aberration of light. It would occupy too much time to go fully into this, and I must content myself with giving you a general notion of it.

Suppose a person out on a perfectly calm day when rain was falling, and accordingly, on account of

the perfect calmness supposed, falling vertically. If
the person were at rest, he would deem the rain to be
falling vertically. But suppose he were carried horizontally along with a motion so smooth that he was
unconscious of it. The rain though falling vertically
would *appear* to him to fall in a somewhat slanting
direction, as if it came from a point not exactly
in the zenith, but displaced from it towards the
point towards which the observer is being carried.
And if instead of falling vertically the rain be
falling in a slanting direction, it will appear to the
observer, unconscious of his own motion, to slant
differently; in fact, if we compound the velocity of
the falling rain with a velocity equal and opposite
to that of the observer, the direction of motion of
the rain will *appear* to be that of the resultant
velocity.

Now Bradley found that just the same thing takes
place with regard to light. The earth in revolving
round the sun moves at the rate of about 20 miles in
a second, towards a point in the heavens lying in the
plane of the ecliptic 90^0 in advance of the heliocentric
position of the earth, or rather what would be 90^0
if the earth's orbit were strictly circular. The light
which comes from any particular star appears to come
from a place deviating from the true place of the star
towards the point of the heavens towards which the

earth is moving. And as this point goes round the ecliptic in the course of a year, the apparent place of the star describes annually a closed curve, in fact, a small ellipse, round the mean position of the star. The *law* of the apparent displacement of the star is found to be what it ought to be on the supposition that the cause of it is what has been above explained, and the coefficient of the displacement gives the ratio of the velocity of light to that of the earth in its orbit. We are thus furnished with a second means of determining the velocity of light, and observation will show whether the two do or do not agree. The result is a remarkably close agreement when we consider on the one hand the difficulty of fixing on the exact moment of disappearance of a satellite which enters the shadow of Jupiter, and on the other the smallness of the displacement which constitutes aberration, amounting at a maximum to only 20 seconds and a quarter of angle, about the angle subtended by a six-inch rule at the distance of a mile.

It is to be noted that the unit of length in terms of which the velocity of light in both of these methods is primarily expressed is the radius of the earth's orbit, and in order to translate it into miles per second we require to know the solar parallax.

While I am on this point I may mention that in

the year 1849 Fizeau, by an extremely ingenious combination of apparatus, succeeded in determining for the first time the velocity of light by direct experiment. The resulting value was confirmatory of the two astronomical determinations, but was not at first considered as capable of being put in competition with them for accuracy, and no wonder, since the whole interval of time to be measured in Fizeau's experiment, under the circumstances in which it was performed, amounted to only about the one twenty thousandth part of a second. Subsequent determinations, however, first by Foucault, by a different method, and afterwards by Cornu, by Michelson, and by the late Dr James Young and Prof. G. Forbes, by one or other of those methods modified, have rendered the determination so certain and accurate that it is probably quite equal in accuracy to the astronomical determinations. And since the experimental determination gives the velocity in terms of a known distance on the earth's surface, and accordingly in miles or kilometres per second, if we assume the experimental and either of the astronomical determinations as separately valid and sufficiently accurate, by comparing the two we can determine the radius of the earth's orbit, which fixes the scale of the whole solar system, in miles or kilometres. And as we know accurately the dimensions of the earth, we can thus determine the solar parallax, by combining accurate measurements of an astronomical phenomenon frequently recurring or constantly going on with a laboratory determination made once for all.

I have been tempted into a digression by the interest of this subject, and I would now resume the consideration of some of the elementary properties of light, with a view of showing how we are gradually led to the formation of a theory, now thoroughly tested, as to the nature of Light itself.

The observations to which I have last alluded show that light, whatever it may be, takes time to travel, so that in speaking of light as an influence that proceeds *from* a source of light there is no longer occasion to exclude the idea of motion of some sort. I have compared and contrasted two influences of which I have spoken, light and gravitation, and in this last point again it is a contrast with which we are presented; or at least if the two are analogous the analogy has never yet been established. After having ascertained that light takes time to travel, and contemplated that exceedingly curious phenomenon relating to light, aberration, the question naturally presents itself to the mind, Is there anything analogous as regards gravitation? Does it like light take time to travel, or is it an instantaneous influence? Now the consequences which would follow as regards the motions of the bodies of the solar system if gravitation like light took time to travel have been calculated, and it has been concluded that if the influence of gravitation takes time to travel, it at any rate travels incomparably more quickly than light, with a velocity accordingly which is incomparably greater than 186,000 miles per second.

I have reduced to a minimum the mention I have made of the fundamental properties of light, in order not to weary you by repeating what is to be found in every text book, and we may now attack the question, What notion are we to form of the intrinsic nature of this agent, so important to our well-being, so wonderful in the scale of magnitude of the quantities it brings before us, as we have seen already as regards the velocity with which it is propagated, and shall see later on in other respects?

There appear to be but two modes possible of conceiving of a *mechanical* influence emanating from an influencing body, travelling with a finite velocity, and ultimately influencing another body at a distance. A mechanical influence implies the action of matter of some kind, and this matter we may suppose to have been either darted forth, in the manner of a projectile, or to have previously existed in the space between the influencing and the influenced body, to have been disturbed by the influencing body, and then to be successively agitated in successive portions, each portion being agitated by its predecessor, and in its turn yielding to its successor the disturbance so received. We have not far to go to find illustrations of both these kinds of action. Bullets exemplify the first; the progress of waves at the surface of water illustrates the second. A still better illustration, except that it is not visible, is afforded by the phenomena of sound.

Such accordingly are the ideas which lie at the base of the two theories as to the nature of light which for a long time divided the scientific world between them, the corpuscular theory, or theory of emission, and the theory of undulations.

Prima facie there is much to be said in favour of the theory of emission. It lends itself at once to the explanation of the rectilinear propagation of light, and the existence of rays and shadows. It falls in at once with the law of aberration. The laws of reflection and refraction admit of an easy explanation in accordance with it; at least if we except the existence of both reflection and refraction; for according to this theory we should rather have expected beforehand that light would have been *either* reflected or refracted, according to circumstances, not that incident light should have *divided* into a portion reflected and a portion refracted.

The theory of undulations on the other hand presents at the outset considerable difficulties. In the first place it requires us to suppose that the interplanetary and interstellar spaces are not, strictly speaking, a vacuum but a plenum ; that though destitute of *ponderable* matter they are filled with a substance of some kind, constituting what we call a *medium*, or vehicle of transmission of the supposed undulations. When I speak of this medium as a substance, or as material, I mean that it must possess that distinctive property of matter, inertia; that is to say a finite time must be required to generate in a finite portion of it a finite velocity. The necessity of thus filling space with substance seems to have presented a serious difficulty to some minds. In the course of a conversation with Sir David Brewster, who had just returned from France, where he witnessed the celebrated experiment by which Foucault had just proved experimentally that light travels faster in air than in water, I asked him what his objection was to the theory of undulations, and I found he was staggered by the idea *in limine* of filling space with some substance merely in order that "that little twinkling star," as he expressed himself, should be able to send its light to us.

I cannot say that this particular difficulty is one which ever presented itself as such to my own mind.

To me the difficulty is rather that of conceiving such
an influence as that of gravitation to extend across
an absolute void.   Such was the feeling of the great
discoverer himself of universal gravitation.   In a
letter to Bentley, quoted by Faraday as falling in
with his own views, Newton thus expressed himself :—

"That gravity should be innate, inherent and
essential to matter, so that one body may act on
another at a distance through a *vacuum*, without the
mediation of anything else, by and through which their
action and force may be conveyed from one to another,
is to me so great an absurdity, that I believe no man
who has in philosophical matters a competent faculty
of thinking, can ever fall into it.   Gravity must be
caused by an agent acting constantly according to
certain laws ; but whether this agent be material or
immaterial, I have left to the consideration of my
readers."

If the supposition that light consists in undulations
obliges us to suppose that space is filled with some
kind of substance, at least as far as the remotest star
that our most powerful telescopes reveal to us, may
it not be that that same substance forms, in some
manner as yet unknown to us, the link of connexion
whereby the sun is enabled to attract the earth, and
keep it in its orbit ?   It is true that notwithstanding
the labours of various scientific men we are not in a

condition to give an explanation of gravitation, but our inability to explain it by no means proves that it is a primary property of matter, incapable of explanation, or forbids us to suppose that it may in some way be brought about through the intervention of that same substance which we find necessary to assume for the explanation of the phenomena of light on the theory of undulations. And it is quite conceivable, and we may now say even probable, that this same substance has yet other offices to fill. Perhaps the most remarkable of all the investigations of the late Professor Clerk Maxwell is that in which he showed that there is a certain velocity, numerically determinable by purely electrical experiments which can be made and have been made in the laboratory, and expressing the velocity of propagation of an electrical state, which is identical with the velocity of light within the limits of error of the experiments and observations whereby the two have been determined. Assuming for the moment, as a thing at the present day resting on evidence quite overwhelming, that light consists of undulations, we cannot fail to be impressed by the multiplicity of purposes, all bearing so intimately on our well-being, which it seems probable, or not unlikely, are fulfilled by one and the same substance, endowed with properties which we are only gradually learning.

S.   2

I have ventured to allude for a moment to the present state of the theory of light, and I will now go back. The necessity of assuming the existence of some kind of substance in what we commonly speak of as a vacuum, does not appear to have been a serious preliminary difficulty in the way of the reception of the theory of undulations. A far more formidable difficulty appeared at first to be presented by the existence of rays and shadows. It was this that led Newton to adopt the theory of emission, though even he was led in the course of his researches on light to suppose that there was some sort of medium through which the particles of light moved, and in which they were capable in certain cases of exciting a sort of undulation. But the supposition of particles darted forth seemed to him necessary to account for shadows. If light consisted simply of an undulation propagated through a medium of some kind filling the interstellar spaces, as we know sound consists of an undulatory movement propagated in the air, how can we conceive of the existence of shadows, knowing as we do that sound passes freely round corners, and diverges after passing through apertures, though indeed it is true that the freedom is not absolute? Newton's contemporary Huygens was more bold, and adopted the theory of undulations pure and simple, rejecting altogether the notion of particles darted forth from the luminous body, and travelling with the velocity of light. Huygens made a grand attempt to explain the existence of rays, the great stumbling-block at the threshold of the theory of undulations. The principle which goes by his name lies at the very foundation of the theory of undulations, and itself rests on a strictly mechanical basis, being in fact merely an application to the particular question under consideration of the general dynamical principle of the superposition of small disturbances. But this principle does not by itself alone suffice for the explanation of rays. It proves, or at least appears to prove, too much. It is as applicable to sound as, on the supposition that light consists in undulations, it is to light; and if Huygens's explanation of rays were complete there ought equally to be rays of sound, and sound ought to present the same sharp shadows as light.

Huygens attempted to get over this difficulty by entering on certain speculations as to the ultimate constitution of the *ether*, as we call the supposed medium which is the vehicle of light, and as to the mode of action, one on another, of the ultimate molecules of which he imagined it to consist. In this he abandoned the simplicity of the fundamental conceptions of the theory of undulations, and adopted a mode of reasoning not strictly allowable. For the transmission of regular undulations, of which the period is arbitrary, at least within wide limits, requires us to suppose that the transmitting medium is either continuous or may be treated as such; that if it consist of ultimate molecules, or be otherwise heterogeneous, the number of intervals from molecule to molecule, or of deviations of one sign or another from an average homogeneity, shall be very great and as good as infinite within the length of a single undulation; and we have no right to extend to the medium treated as a whole, and regarded as continuous, a mode of communication of motion applicable only to the communication from one to another of a set of discrete molecules.

Accordingly, notwithstanding all that Huygens has done, the existence of rays and shadows, one of the most obvious properties of light, had received no satisfactory explanation on the theory of undulations such as it came from the hands of Huygens; and in this condition it remained for considerably more than a century. His explanation of the laws of reflection and refraction leaves nothing to be desired, except in so far as these laws involve the conception of rays. I cannot now speak of his discovery of the laws of double refraction in Iceland spar, because it belongs to a different branch altogether of the subject.

The theory of rays and shadows long remained in this unsatisfactory state; in fact, till quite the end of

the last century. Newton's discovery of the compound nature of white light showed that there must
be in light an element of some kind susceptible of
continuous variation. Each theory, the corpuscular
and the undulatory, furnishes elements susceptible of
continuous variation. What the element is, on the
theory of emission, has not, so far as I know, been
specified by the supporters of that theory, and difficulties seem to attend whatever supposition in that
respect you can make. The theory of undulations
presents one, and I may say but one, element which
might serve for the purpose, namely, wave length, or
what comes to the same thing, periodic time. That
in fact periodic time must be the element variations
in which correspond to variations in refrangibility, is
clearly pointed out by other phenomena which I have
not as yet touched upon. But this development of
the theory did not take place till the present century,
though some of the leading facts on which it is based
were known to and studied by Newton. Accordingly
from the time of Newton till the end of the last
century, and even further, the theory of emission was
that chiefly in vogue with scientific men. Various
causes probably contributed to this result. The
rectilinear propagation of light at first sight looks
more like the motion of projectiles than the propagation of undulations, which in cases of what are

undoubtedly undulations spread out much after being laterally confined. The impetus given to the study of the motion of particles under the action of known forces by Newton's great discovery of universal gravitation, turned the labours of men of science into that channel rather than to a study of the propagation of vibrations. The great weight again of Newton's authority had doubtless its share in leading men to follow the theory as to the nature of light which he had taken up.

It is probably due to this preponderating influence of the theory of emission that so little notice was taken of the theory of the aberration of light. In the explanation of the phenomenon which is contained in the ordinary text-books of astronomy, which has doubtless descended traditionally from that given in earlier treatises, it is quietly assumed as a matter of course that the rectilinear propagation of the light coming from a heavenly body is not disturbed by the motion of the earth. Did light consist of particles darted forth, there is no reason to suppose that it should; in fact, to make such a supposition would be to fly in the face of all we know respecting the action of attracting forces, since any motion of the attracting body does not enter into account. But on the theory of undulations it is far otherwise. We should naturally have been disposed to look on the

earth in its motion round the sun as ploughing its
way through the ether. Now if light consist of undulations propagated through this ether, we might
have expected that the ether being pushed by the
earth out of its way, the course of the undulations
which it carried would be affected, possibly in an
irregular way, in case eddies were produced, and at
any rate in a manner which there appears no reason
a priori should be in conformity with the simple law
of aberration.

Accordingly Dr Young, to whom mainly we owe
the revival of the theory of undulations which took
place about the beginning of this century, supposed
that instead of the earth's pushing the ether out of its
way, it allowed it to pass freely through its substance,
8000 miles though it be in thickness, far more freely
than a grove of trees transmits the wind ; and that in
consequence of this perfect freedom of passage, the
ether outside the earth's surface was not disturbed by
the earth's motion, nor consequently the undulations
passing through it.

Now startling as is this supposition, and contrary
to all that we should have anticipated, we cannot say
that it must be rejected. For we must remember
that we have no direct evidence even of the existence
of an ether; it is not directly recognizable by any of
our senses ; its properties may be, and doubtless are,

very different from those of ponderable matter, and we must be content to learn them by degrees, as they may be revealed by the study of the phenomena which are referable to actions of the ether. Nevertheless we are not absolutely driven to accept Dr Young's hypothesis; for there is as I have shown another way in which the law of aberration may be obtained; a way which though not free from difficulties exempts us from the necessity of supposing that the earth in its motion through the ether allows the ether to pass through it with absolute freedom.

At this point it may be well to pause for a moment and consider the probabilities in favour of the two hypotheses. The existence of rays and shadows seems perfectly simple according to the theory of emissions; as far as we have gone it presents a serious difficulty on the theory of undulations. The laws of reflection and refraction leave little to choose between the two. In one respect indeed the theory of undulations has the advantage; for it indicates that there ought to be a partition of the incident light into a portion reflected and a portion refracted, whereas on the other we should rather have expected that the light would have *either* been reflected or refracted. This advantage is however of no great weight, for it would not be difficult to frame plausible hypotheses in the theory of emission which would lead to a partition. As regards aberration, the corpuscular theory has a decided advantage, for on it the explanation of the phenomenon is perfectly simple, whereas according to the theory of undulations all we can say is that it is not inexplicable.

The balance on the whole seems to lean towards the side of the corpuscular theory. And yet that theory is now altogether exploded, and the rival theory is established on so firm a basis that no one who has studied the subject can doubt that the second of the two modes of conception with which we started expresses the truth, and that light really consists of a change of state propagated from point to point in a medium existing between the luminous body and that which the light affects.

It may be said, If the former theory is now-a-days exploded, why dwell on it at all? Yet surely the subject is of more than purely historical interest. It teaches lessons for our future guidance in the pursuit of truth. It shows that we are not to expect to evolve the system of nature out of the depths of our inner consciousness, but to follow the painstaking inductive method of studying the phenomena presented to us, and be content gradually to learn new laws and properties of natural objects. It shows that we are not to be disheartened by some preliminary difficulties from giving a patient hearing to a hypo-

thesis of fair promise, assuming of course that those
difficulties are not of the nature of contradictions
between the results of observation or experiment and
conclusions certainly deducible from the hypothesis
on trial. It shows that we are not to attach undue
importance to great names, but to investigate in an
unbiased manner the facts which lie open to our
examination.

I now come to other classes of phenomena with
respect to the explanation of which there is the widest
possible difference between the two hitherto rival
theories respecting the nature of light. We are
all familiar with the vivid colours of soap bubbles,
colours which are exhibited in the case of transparent
solid plates or liquid films which are excessively thin,
or of very thin plates of air contained between two
surfaces of glass which are in contact, or almost in
contact, at one point. It is to Newton that we owe
the first investigation of the laws of these " colours of
thin plates " as they are called. By placing a convex
lens of small curvature in contact with a plane piece
of glass, we obtain a separating plate or film of air
the thickness of which vanishes at the point of contact,
and increases very slowly at first on receding from it.
In this case the colours are arranged in circles round
the point of contact, the rings forming annuli of
increasing diameter but decreasing width as we

recede from the centre outwards. The centre is dark ; and when the incident light is white, on going outwards about seven alternations can be traced, after which the field is sensibly of uniform illumination and free from colour. The squares of the diameters of the rings are found to increase in arithmetic progression in passing from ring to ring.

When instead of using white light the system of the rings is illuminated by the colours of a pure spectrum, a vast number of rings is seen ; in fact, they go on till from the increasing narrowness of the annuli they become too fine to be seen. In this case the colour naturally remains the same, being that of the part of the spectrum that is used to illuminate the glasses. The scale of the system changes greatly with the colour, decreasing from the red to the violet.

The scale of the system depends very greatly upon the curvature of the lens ; and to form the rings on a large scale, which is convenient for examination, it is necessary to use a lens of small curvature. The thickness of the interposed plate of air where any ring is formed can be obtained by an easy calculation from the diameter of the ring and the radius of curvature of the lens, both which can be measured. When this is done it is found that a given ring is always formed where the plate has a given thickness.

If the interposed medium be water instead of air,

or generally any transparent fluid, the scale of the rings is diminished. It is found that the square of the diameter of a given ring varies inversely as the refractive index of the interposed liquid.

As my object is not to give a complete account of the phenomena of the rings, much less a complete explanation of the phenomena, I shall dwell no further on the appearances which the rings present under different circumstances, the features I have already described being sufficient for my purpose, which is to give a fair idea of the evidence on which rests that theory as to the nature of light which we shall be led to adopt.

Newton endeavoured to account for the various phenomena of thin plates by his celebrated theory of fits of easy reflection and transmission. We have seen that the existence of both reflection and refraction is a difficulty in the theory of emission which has in some way to be accounted for. It will not do to suppose that there are two permanently distinct kinds of light, of which one, when the light falls at a given inclination on a given substance, is always reflected and the other always transmitted, for if either the reflected or the transmitted beam be allowed to fall at the same angle on the same substance, it is divided into a reflected and a transmitted beam. We must therefore suppose, even independently of the phenomena of thin plates, that the same particles of light are sometimes in a condition to be reflected and sometimes in a condition to be transmitted : and we have only further to postulate that these changes of state, whatever may be their nature, take place with a regular periodicity in order apparently to account for the phenomena of thin plates, at least if we restrict ourselves to the case of a perpendicular incidence on a plate of given kind, such as a plate of air between two surfaces of glass. For all the light which is transmitted by the first surface must be in a fit of easy transmission which it does not at once lose, so that if it falls immediately on the second surface it is transmitted, and therefore the central spot looks comparatively black. The same must be the case where the distance between the surfaces is equal to the length of a fit, or 2, 3, &c. times the length of a fit ; and for intermediate thicknesses of the plate, corresponding to intermediate distances from the point of contact, the effect will be intermediate, and there will be more or less reflection, which will be greatest for thicknesses exactly half way between the critical thicknesses above mentioned.

But tempting as this theory at first sight appears, though if it be true it must be left to subsequent research to indicate what that periodic element in relation to the particles of light can possibly be which shows itself by an alternate capacity for reflection and refraction, it fails completely to account for the other features of the phenomenon, some of which I have mentioned. We should have expected beforehand that the length of a fit would have been independent of the angle of incidence ; and when we learn that to reconcile theory and observation we must *suppose* it to vary as the secant of the angle of inci-dence, we see no way of accounting for such a law.

Again, as to the effect of a change of medium, the most natural supposition to make would be that the constant element of periodicity which characterises light of any particular refrangibility is a constant periodic time. Now the explanation of the law of refraction according to the theory of emission requires us to suppose that light travels faster in refracting media than in vacuo, in the ratio of the refractive index to unity. We might have expected accordingly that the length of a fit in water would be greater than in air in that ratio. We have seen however that it is less in the inverse ratio. We are unable to frame any plausible hypothesis, on the theory of emission, why it should be so.

Accordingly the theory, if such it can be called, consists merely of a set of incoherent laws, not indicated beforehand by theory, not even falling in with it after they have been pointed out by obser-vation, and it has accordingly nothing about it which seems to bear the stamp of truth.

Nor is this all. Even as regards the formation of the rings at a perpendicular incidence, where at first sight the theory appears to be most successful, it leads to a conclusion which is belied by observation. According to this theory, the office of the first surface of the plate is solely one of sifting, by reflecting back those particles of light which are in a fit of reflection, and thereby preparing the way for the alternate transmission and reflection of the particles at the second surface. There should therefore be as much light reflected at the first surface as if the second glass were away altogether, and therefore the dark rings should be only comparatively dark. The experiment may be easily and successfully tried with homogeneous light, such as that of a spirit-lamp with salt on the wick, and it is found that the dark rings are decidedly darker; in fact as to sense apparently black.

LECTURE II.

Interference—Explanation of the colours of thin plates afforded by the theory of Undulations—Diffraction.

IN my last lecture I pointed out the insufficiency of the theory of emission to account for the various phenomena of the colours of thin plates. Let us now see whether the theory of undulations lends itself to an explanation. In the first place the element of periodicity, instead of being wholly extraneous to the fundamental idea, and hardly if at all to be reconciled with it even when suggested by phenomena, is one naturally, almost inevitably, involved in the fundamental conception. I say, *almost* inevitably; for a set of undulations *might* consist of a succession of isolated pulses following one another in a wholly irregular manner; but the analogy of sound would make it far more likely *a priori* that they should form a series of regularly periodic disturbances; and the supposition that such is their character therefore harmonizes perfectly with what we should have expected beforehand.

Supposing then that the undulations we have to deal with are regularly periodic, can we on the undulatory theory give any account of the alternations of light and darkness which we observe in Newton's rings?

The explanation which this theory affords is based on a very simple and very general dynamical principle, of very wide application, called the principle of the superposition of small motions. For the sake of those who may not have much attended to dynamics, I will endeavour to give some idea of what this principle means.

Suppose a stone thrown into still water. It produces as we know a series of small waves, which spread out in the form of circles from the place of disturbance. What passes outwards is, not the material particles of the water, but a certain state of things. If we observe a minute floating body, it is seen to retain its average position, and merely to move very slightly backwards and forwards, up and down. So far, we have merely a visible illustration of the progress of an undulation, differing it is true from those belonging to sound or light in the fact that it is only in the neighbourhood of a particular surface that the disturbance is sensible, which does not however prevent it from forming a useful illustration for those to whom the conception may be new.

But now instead of a single stone, suppose that

there are two similar stones thrown in simultaneously at a little distance apart. Each will give rise to a series of circular waves which at first will be distinct, each diverging from its own centre, but not yet having reached one another. But presently the waves from the one centre will invade the region occupied by those from the other. What will then take place? According to theory, each disturbance will find the mass in which it is being propagated under as nearly as possible the same conditions, so far as itself is concerned, as if the other disturbance did not exist. The consequence is that any particle of the mass will be as much displaced from its mean position as if the other disturbance did not exist, and its actual place will therefore be found by compounding, as it is called, the displacements due to the two disturbances taken separately. Accordingly where a ridge due to one of the disturbances coincides in position with a ridge due to the other, we have an elevation of double height, supposing we choose a place where the intensities of the disturbances due to the two sources are the same ; but where a ridge due to one falls in with a trough due to the other, we have neither elevation nor depression, but the water is found at its natural level.

Now if there be any truth in the theory of undulations, something of the same kind must take place with light. If we have two sets of disturbances always exactly alike, then when the conditions are such that the two disturbances separately considered take place always in opposite directions, being superposed they simply produce a disturbance *nil*, whereas when the directions agree they produce a disturbance of greater amount than either separately. And if light consist of such a disturbance in the ether, then in the former case we ought to have no light, whereas in the latter the light ought to be greater than if the one set of disturbances alone existed. And if the two sets of disturbances, while still in other respects alike, differed in amplitude in a constant ratio, the only difference that would make in the result would be that the light would not vanish at the minima.

But it must be particularly remarked that in order that this neutralization should always take place in the same way at the same place, it is essential that the two sets of disturbances should always agree as to their times of starting, or differ, if they do differ, by a perfectly constant quantity. For if one were half an undulation ahead of the other, that would make all the difference whether the two disturbances strengthened or neutralized each other: and if the relative times of starting varied irregularly a vast number of times in a second, we should have at a given place neutralization and cooperation succeeding one another so rapidly that nothing but the mean effect would be perceived, and that, as may be shown, would be simply the sum of the mean effects taken separately. Now if the disturbances came from two independent sources, such as two different portions of a flame, the relative starting points of the two would be purely casual, and no fixed and permanent neutralization would be to be expected. And obser‑ vation shows that none of those alternations of light and shade which on the theory of undulations we refer to interference are manifested unless the two interfering streams of light have come originally from the same source, having subsequently pursued slightly different paths.

Let us now see how these principles apply to the explanation of Newton's rings. If we consider a small portion of the thin plate of air by reflection from which they are seen, we perceive that there are two reflecting surfaces near one another, and conse‑ quently two reflected streams, one reflected from the upper surface of the plate of air, that is, internally in the upper glass, and the other reflected from the under surface of the plate of air, that is, from the upper surface of the under glass. These are it is true accompanied by other streams which have been reflected backwards and forwards internally in the plate, so that the total number of reflections in these streams is 3, 5, 7, 9...; but these streams, being usually comparatively weak, may in a general explanation be left out of consideration. Now the second of these streams has had to travel a very little further than the first. In the simplest case, to which we may confine ourselves for the present, that of a perpendicular incidence, the excess of length of path of the second stream is evidently just double the thickness of the plate of air, and varies accordingly from point to point of the field of view, the thickness increasing slowly at first and afterwards rapidly as we recede from the point where the lenses are in contact. Close to the point of contact, the second stream is only imperceptibly behind the first, and the two might be expected to be in accordance, that is, to agitate the ether in the same direction, and therefore to strengthen one another. As we take a point further outwards from the point of contact, the retardation increases, and when it becomes half the length of a wave the two would be in opposition, that is, one would agitate the ether in a way just opposite to the other, and therefore, supposing the intensities of the two the same, as would nearly if not exactly be the case, they would neutralize each other, and darkness would be the result. And as the thickness of the plate of air between the two glasses varies as the square of the distance from the point of contact, the law of increase of the rings for any one kind of light as the order of the ring increases follows at once, and moreover the law that under different circumstances as to curvature of the glasses the same ring is always formed where the thickness of the plate is the same.

But there is one marked point of contrast between the results of theory, so far as I have yet explained it, and observation, namely, that we have been led to expect a maximum of brightness where the paths of the two streams are the same, that is close to the point of contact, or where they differ by one, two, or a complete number of undulations, and a maximum of darkness if not absolute blackness where the difference is an odd number of half undulations. This would give the places of light and shade exactly reversed as compared with observation, which would show one of two things, either that the theory must be rejected, or that some circumstance has been overlooked which, had it been taken into account, would have made the theoretical result right in this respect.

Now the reflections which the two streams respectively undergo take place under very different circumstances. The first stream is reflected internally in glass when the light, which had previously been travelling in the glass of the upper lens, arrives at the under surface of the glass, or upper surface of the interposed plate of air, the second is reflected back into air at the upper surface of the under lens.   As the two reflections take place under different, in some respects opposite, circumstances, there seems nothing unlikely *a priori* in the supposition that the signs of the reflected vibration should be opposite in the two cases.   Dynamical analogies are not far to seek ; for example, we know that when sound travelling along a pipe meets a closed end, it is reflected in such a manner that condensation in the reflected answers to condensation in the incident, and rarefaction to rarefaction ; but when it is reflected, as reflected it is, from the open end of a tube, rarefaction answers to condensation and condensation to rarefaction.   And this theoretical conclusion as to the opposition of signs in the two cases, so probable from analogy, is converted into a certainty by the application of a simple dynamical law of great generality, which may be called the law of reversion.   Hence then the occurrence and features of the rings at a perpendicular incidence, for any one kind of light, are on the undulatory theory matters of pure prediction.

Next consider the effect of inclination.   It follows from a very easy calculation that the effect of the thickness of the plate of air at any inclination in producing retardation in the stream reflected from the second surface of the plate, relatively to the stream reflected from the first surface, is the same

as it would be at a perpendicular incidence for a
thickness smaller in the proportion of the cosine of
the inclination of the light while in the plate of air to
unity, or, which is the same thing in the case of a
lens with nearly parallel surfaces for the upper glass,
of the cosine of the angle of incidence on the first
surface to unity. Hence follows at once the simple
law of dilatation of the rings on increasing the angle
of incidence, namely, that the square of the diameter
of any ring varies as the secant of the inclination on
the first surface. Of this law, as we have seen, the
theory of emissions could give no account.

Again, the effect of the substitution of water for
air between the lenses in causing the rings to con-
tract, and the law of their contraction already mentioned, follow immediately on this theory from first
principles, since the explanation of refraction on the
theory of undulations necessitates the supposition
that light travels more slowly in refracting media
than in vacuum, in the proportion of the refractive
index to unity.

In short, the theory completely explains the phenomena of the rings seen by reflection. And this is
true even as regards the more minute features, into
which I have refrained from entering. The New-
tonian "fit" at a perpendicular incidence expresses
half the length of a wave; and as the scale of the

rings decreases from the red to the blue, and accordingly as the refrangibility of the light increases, we learn that the variable element in light which corresponds to a change of refrangibility and change of colour must be the wave-length *in vacuo*, or, what comes to nearly the same if not exactly the same thing, the periodic time. The compound tints of the rings are explained here as in all other optical phenomena, whatever theory of light we may adopt, by the superposition of the ring-systems corresponding to the different kinds of light of which white light consists, which are on different scales as to size; and the fact that not more than about seven rings are seen with white light merely depends on the overlapping of the rings of high orders corresponding to the different colours.

I have hitherto said nothing about the system of rings in the transmitted light, which are complementary in character to the rings seen by reflection, but far less vivid, resembling in fact a vivid system complementary in character to the reflected system, overlaid by a comparatively large quantity of uniform white light. In nearly every respect the theory of these is analogous to that of the reflected system, so that whatever theory explains the one can hardly fail to explain the other. There is just one feature the explanation of which involves considerations into which I have not entered, and to this I will confine myself.

The transmitted system, according to the explanation afforded by the theory of undulations, depends on the interference of two streams of light, one passing right through the plate of air comprised between the lenses, and the other following it after two reflections in air at the adjacent surfaces of the glasses. These are associated with other portions which have been reflected 4, 6 ... or any even number of times; but these, being usually comparatively weak, may in a general explanation be disregarded.

Now when light is incident perpendicularly, or nearly so, on the surface of crown glass, about 4 per cent. of the incident light is reflected, and the rest enters the glass. The intensity of light once reflected in this manner being thus only the $\frac{1}{25}$th of the intensity of the original, the intensity of light twice reflected will be only the $\frac{1}{25}$th of that, or the $\frac{1}{625}$th of the original. Now as the light which passes through one surface loses only 4 per cent. by reflection, the light which passes through the two surfaces of the plate still retains the 0.96^2 or 0.9216 of its original intensity; so that the twice reflected light has an intensity only the $\frac{1}{676}$th of the direct, hardly more than one-sixth per cent. How then, it might be said, could so small an addition or subtraction of light be perceptible at all, and so produce in homogeneous light differences of intensity, and with white light changes of colour, which though not it is true by any means so striking as in the reflected system of rings are nevertheless very evident, and demand far greater differences of intensity than that ? ·

The explanation of this paradox lies in a consideration of the relation between intensity and amplitude of vibration. If light consists in a disturbance of a subtile medium, or ether, then the greater be the disturbance, other circumstances being the same, the stronger must be the light. But supposing the amplitude of excursion of the ether to be doubled, trebled... is the intensity of the light doubled, trebled ... or if not, in what other proportion is it increased ?

Now a number of different considerations lead decisively and independently to the same conclusion, namely, that the intensity is to be measured by the square of the amplitude of excursion. Hence if the amplitude of excursion to and fro of the ether is increased in the proportion of 1 to 2, 3..., the intensity of the light is increased in the proportion of 1 to 4, 9...

Suppose now that two disturbances from the same source, and following nearly the same path, have a, b for the coefficients of excursion due to them sepa-

rately; when acting together the coefficients of excursion will fluctuate between the limits $a + b$ and
$a - b$ according as the phases are in agreement or in
opposition. Hence whereas the intensities of the two
separately will be in the ratio of a^2 to b^2, the maximum
and minimum intensities due to the compound disturbance will be proportional to $(a + b)^2$ and $(a - b)^2$.

Suppose now the first disturbance to be very
much greater than the second, then a will be very
much greater than b. Hence whereas the sum or
difference of the intensities of the two streams would
be got by adding or subtracting b^2 to or from a^2, the
maximum and minimum intensities of the compound
disturbance will be got very nearly by adding or subtracting $2ab$, a quantity which is greater than the
former in the proportion of $2a$ to b. Thus with the
numerical values just mentioned, which are applicable
to Newton's rings at a perpendicular incidence, if we
take the coefficient of vibration and the intensity of
the stronger stream each for the unit of their respective kinds, the coefficient of vibration of the weaker
stream will be $\frac{1}{25}$ and the intensity $\frac{1}{625}$ nearly, whereas the maximum and minimum intensities of the
compound stream will be $1 \pm \frac{2}{25}$ nearly, and the difference between them will be as great as $\frac{4}{25}$.

In the particular case of Newton's rings, the
weaker stream is not readily viewed apart; but in

certain experiments of diffraction, a subject that
will be touched on by and by, the two streams lend
themselves readily to separate observation, and the
occurrence of distinct fringes of interference theoretically referable to an invisible agent—invisible because too faint to be seen—is not a little paradoxical
in appearance.

It appears then that the fundamental hypotheses
of the undulatory theory suffice to account in the
most complete manner for the phenomena of Newton's rings, and the colours of thin plates in general,
without making any fresh assumption whatsoever.
It is to Dr Young that we owe the explanation of
these colours on the theory of undulations, an explanation given at a time when any other theory than
the corpuscular could hardly gain a hearing : and to
him also we owe the first direct experiment proving
that certain fringes which connect themselves by
numerical relations with Newton's rings are incon-
testably due to interference.

As introductory to this experiment, it will be
convenient to mention another phenomenon which
forces itself upon the notice of the observer simultaneously with that with which we are more immediately concerned, and which therefore it is desirable to
be able at once to refer to its proper place, besides
that at a more advanced stage of our study of the

subject it will be found to be of very great importance.

Suppose that the sun's light is reflected horizontally into a darkened room, passing through a
very small hole in the shutter, or what is more convenient through a lens of short focus. Let the light
be allowed to fall on an opaque screen at the distance
say of a few feet from the luminous point, the screen
being terminated by a straight edge, suppose vertical.
Let the light passing the edge of the first screen be
received on a white vertical screen which we may
suppose a few feet further off from the luminous
point. According to geometrical optics, if we project
the edge of the opaque screen on to the receiving
screen, by straight lines drawn from the luminous
point, all to the illuminated, suppose the right-hand,
side of the projection we shall have uniform illumination, the same as if the opaque screen were away,
while all to the left of it we shall have darkness.
According to observation, there is no such abrupt
transition on the receiving screen from uniform darkness to uniform brightness. The illumination increases continuously, though rapidly. The illumination begins to be sensible before we reach the
geometrical shadow, or projection of the straight
edge, where it is still considerably feebler than the full
illumination at a distance on the right. On going in

a right-hand direction from the geometrical shadow the illumination rapidly increases, and actually becomes considerably greater than the full illumination at a distance. It then decreases to a minimum, increases to a maximum, and so on, the maxima and minima differing less and less, by excess and defect respectively, from the illumination at a distance, and gradually occurring in more rapid succession, till they become insensible. Three such bands can usually be traced before the illumination becomes sensibly uniform. With white light the scale of the bands, and of their distances from the geometrical shadow, changes as usual from colour to colour, decreasing from the red to the blue. The appearance is the same at all distances of the two screens from the luminous point, the scale, merely, of the system varying according to circumstances, and likewise the smallness demanded in the source of light in order that the bands may not be confused. From the constant character of these bands we readily recognise them in experiments in which they appear associated with other phenomena.

Suppose now that instead of the opaque screen extending indefinitely on one side, the light is intercepted by a narrow slip, a knitting-needle, or anything of the kind. The shadow is bounded on both sides externally by fringes of the character of those

just described, and which accordingly we must attri-
bute to the same cause, referring the right-hand set
to the light which passed to the right of the obstacle,
acting independently of that which passed to the left,
and similarly as regards the external fringes of the
left.   But besides these the shadow itself is occupied
by another set of coloured fringes, finer usually than
the former, and unlike them of equal width throughout.   They may be viewed through an eye-lens; and
thereby magnified. If the screen be semi-transparent,
we may view them with the lens from behind, and
now if we take away the screen altogether, using
nothing but the lens, and receiving the light directly
into the eye, we see them as before.   But now they
are much brighter, since the light is no longer scat-
tered in all directions by a screen, so that we can
afford to use a narrower source of light, suppose a
lens of shorter focus in the window; and having the
narrower source we may use a wider opaque slip
without the fringes getting confused. Under these
circumstances the system of internal fringes occupies
only the middle portion of the shadow, being well
separated from the external fringes on both sides.
The very centre of the shadow is bright for all the
colours of the spectrum, and the middle bright band
is accordingly white, except just at the edges, where
it is slightly reddish, but after that the bands right

and left soon become coloured, on account of the difference of scale for the different colours.

Now what account can we give of the formation of these internal fringes on the theory of undulations? We have seen that when light passes by an opaque screen, it does not at once become insensible within the limits of the geometrical shadow; a little light bends round the edge into the darkness, enough to produce a feeble and rapidly decreasing illumination on a screen placed to receive it. This "inflexion" of light is a fact, account for it as we will: and if light consists of undulations, the analogy of those undulations which we can directly examine would lead us to infer that light must be inflected into a shadow. Indeed the grand original difficulty which for so long a time prevented the reception of the theory of undulations was that of accounting for the comparative absence of inflexion; in other words, for the existence of rays and the sharpness of shadows. Admitting this small inflexion as a fact, we see that at any rate we can not be far wrong in supposing the bending to take place close to the edge of the obstacle. On this supposition, the length of path of the inflected stream in travelling from the luminous point to the point of the field, or focal plane of the eye-lens, where we seek the illumination, will equal the path from the luminous point to the edge of the

obstacle, *plus* the distance from thence to the point in
the field. The difference of paths for the two inflected streams which reach the same point of the
field is accordingly easily found, especially as we are
only concerned with points lying but a very small
distance from the plane passing through the luminous
point and the middle of the obstacle, or say the
central plane. The difference of path vanishes at the
central plane, where accordingly we have a maximum
of brightness, and on receding in a lateral direction
from the central plane the difference of path changes
in proportion to the lateral distance, and accordingly
it is equal to 1, 2, 3... wave-lengths for a series of
equidistant straight lines parallel to that in which the
central plane cuts the field. These lines are what
ought according to theory to be the middle lines of
bright fringes, and lines midway between them ought
to be the middle lines of dark bands separating the
fringes. The theoretical breadth of a fringe, being
sensibly independent of the distance of the luminous
point, is connected with the wave-length by a very
simple formula, involving only the distance from the
opaque slip to the screen on which the light is received
and the breadth of the slip, both of which can easily be
measured. On the other hand, the fringes are actually
seen, and their breadths can be measured. Comparing the theoretical and observed breadth of a

fringe, we obtain the value of the quantity which if the theory be true expresses the length of a wave. On comparing this with the measure obtained from Newton's rings, we obtain identically the same value within the limits of errors of observation.

This numerical relation indicates that the alternation of light and dark in these two different phenomena is referable to a common cause, whatever that may be. Newton endeavoured to explain the rings which go by his name by the theory of fits of easy reflection and transmission. We have seen that this theory failed to account for the dilatation of the rings produced by increasing the angle of incidence, and for their contraction produced by the substitution of water for air as the interposed medium. Still, it gives correctly the law of increase of the radii of the rings at a perpendicular incidence as the order increases, and the law connecting the scale with the curvatures of the lenses. Now confining ourselves to the case of a perpendicular incidence, where the theory of fits is most successful, we may notice one radical difference between the explanations offered by the theory of fits and the theory of undulations. According to the former, the office of the first surface of the thin plate is simply one of sifting, and it is by reflection or non-reflection from the second surface that the rings are formed. According to the theory of undulations on

4—2

the other hand, there is no sifting at all : the light reflected from and the light transmitted by the first surface have both one and the other just the same properties as the incident light, and it is by the simultaneous working of the light reflected from the upper and that reflected from the under surface that the alternations of illumination are produced. Now in the experiment of the internal fringes, there is absolutely nothing to sift the light, and yet we get alternations of light and darkness as in Newton's rings, and what is more are conducted to the very same measurable length, clearly representing something inherent in the nature of light, which according to the theory of undulations is simply the length of a wave, the conception of which is radically involved in the fundamental points of the theory. We can hardly refuse to admit that the alternations of light and dark witnessed in these two phenomena are really due to interference; to the simultaneous working of two portions of light.

This conclusion was converted we may say into a certainty by a celebrated experiment of Dr Young's on the internal fringes. He showed that if a small opaque screen were placed so as to intercept the light going to fall on one side of the narrow slip, or else which had already passed it, in either case the central fringes disappeared as well as the external fringes on the same side of the slip. And if a plate of glass were substituted for the opaque screen the internal fringes disappeared as before, but now the external fringes were seen on both sides. It is clear therefore that the internal fringes are really due to the joint working of the two portions of light inflected in passing the two edges, while as regards the external, those which appear on the right and left are independently produced by the light which passes on the right and left respectively of the opaque slip. The disappearance of the internal fringes occasioned by the interposition of a piece of glass on one side only of the opaque slip is explained in the same way as the non-exhibition of Newton's rings, when white light is used, outside a very moderate number surrounding the point of contact of the glasses.

After what precedes, the reality of the interference of light might well be taken as established. It is to be noted however that in the theory of the last experiment we have had to take for granted the inflexion of light. It is true that the fundamental conceptions of the theory of undulations lead us to expect inflexion, the difficulty being rather to explain how there is so little. We have not however as yet seen how the determination of the inflexion is to be brought within the domain of theory, and to that extent we have been working in a field not yet fully explored.

It is to Fresnel that we owe the first experiment of interference in which there is neither a thin plate, the first surface of which might according to Newton's views have exercised a sifting action on the light, nor anything unusual, such for example as inflexion, occurring in the progress of the light, but in which the two portions of light that interfere are simply regularly reflected or refracted as the case may be.

This object was accomplished by Fresnel in two ways, by reflection and by refraction: by a pair of interference mirrors, and by an interference prism, which is a prism with a very obtuse angle, such as 179^0. It will be sufficient to mention the former method.

Let two plane mirrors be procured, such as two pieces of plate glass, blackened on the back, each mirror being bounded by a straight edge. Let them be mounted so that the straight edges are close together, and the planes of the mirrors are nearly but not quite continuations of each other; the planes of the faces making with each other a very obtuse angle, suppose only a fraction of a degree less than 180^0, the concavity being on the side of the reflecting faces. Great care must be taken that neither mirror juts out above the other where they meet along the straight edges, which may be ascertained by passing the finger lightly across the junction. Let the sun's rays be reflected into a darkened room, and brought

to a point, or what may be regarded as such, by a lens
of short focus. Let the light proceeding from the
luminous point be received, say at the distance of a
few feet, on the pair of mirrors. According to geometrical optics, the light reflected from each mirror
will proceed after reflection as if it came from the
virtual image of the point in the mirror. If each
mirror be projected into space by lines drawn from
the virtual image belonging to it, we see that there is
a narrow wedge of space within which both reflected
streams mix. If the light be received on a screen
there will be two illuminated areas corresponding to
these two projections respectively, and a narrow more
highly illuminated band where the two overlap, corresponding in fact to a section, by the plane of the
screen, of the wedge above spoken of. If this doubly
bright portion of the field be more carefully scru-
tinized, by removing the screen and receiving the
light directly into the eye through an eye-lens, we
see the field marked on both sides by the usual
external fringes at the boundary of an illuminated
field. But besides these we see, running parallel to
these along the middle of the field, a set of sharply
defined, and commonly much narrower fringes, which
as usual are on a different scale for the different
colours, coarser for the red, finer for the blue. These
are quite different in appearance from the compara-

tively vague external fringes which are seen at both sides of the field. And if any suspicion were entertained that after all these sharply defined fringes, said to be due to the interference of two regularly reflected streams, were really connected with the external fringes, their complete independence would be shown by a very simple modification of the experiment. Adjust the mirrors so that one shall jut out a very little above the other towards one end of the line of junction, and the other towards the other, this will be sufficient to make the line of intersection of the planes of the mirrors altogether oblique to the line of junction. The region of space within which the two streams mix remains as before, and the external fringes which bound it on the two sides, but the sharply defined fringes which before ran along the middle of the doubly bright region, being parallel to the line of intersection of the planes of the mirrors, are now altogether oblique to the line of junction, and accordingly inclined at a considerable angle to the external fringes. When the sharply defined fringes begin to run out of the doubly bright portion of the field, they and the external fringes begin to modify one another, but under the conditions which usually prevail, and at any rate are easily obtained, the former pursue for a long way their oblique course quite undisturbed.

It is proved therefore to absolute demonstration that the property of interference is one essentially belonging to light from its very nature. Two lights of the same kind, that is, of the same refrangibility, from the same source do really strengthen or oppose each other, in the latter case producing darkness if the intensities are equal, according as their lengths of path from the source to where they mix are the same or differ by a multiple of a certain length depending on the nature of the light, or as they differ by an odd multiple of half that constant. And if part of the path lies in some refracting medium, as water, instead of air, it is equivalent, so far as interference is concerned, to a path in air greater in the proportion of the refractive index to unity, and which accordingly would take exactly the same time to be travelled over according to the theory of undulations, since according to that theory the velocity of propagation in media is less than in vacuo in the proportion of unity to the refractive index.

This fundamental constant, which according to the undulatory theory expresses the length of a wave, may be determined more or less accurately by any of the instances of interference above mentioned ; though the progress of our knowledge has furnished us with methods of determining it of still greater exactness. It increases as we have seen, and that considerably, something like in the proportion of 2 to 3, in passing from the blue to the red. As to its absolute value, it will be sufficient to say that for rays of mean refrangibility we may take it in round numbers at the $\frac{1}{50000}$th part of an inch.

We have seen with what admirable simplicity the theory of undulations explains the various phenomena, in all their details, which have been mentioned as referable to interference. And yet the grand original difficulty, that of explaining the existence of rays and shadows, has been left untouched, or received at most only a lame explanation. Yet if the theory be true it ought to be capable of accounting for these phenomena as well as for those, in some respects simpler in character, which have been so successfully referred to interference.

It is remarkable that it was not till the study of the theory of light had made great progress, subsequently to the revival of the undulatory theory about the beginning of the present century, that the elementary phenomena of rays and shadows received their full explanation; and yet, once that explanation is propounded, it is seen to involve nothing more than the very elements of the theory of undulations; to be in fact nothing more than might have been foreseen from the beginning had the human race been sufficiently acute. The history of the explanation compared with our present knowledge affords a remarkable example of the manner in which we start on our course of investigation by mounting on the shoulders of our predecessors ; and so it may very likely be in the future that things that appear to us mysterious, and which we labour hard to explain, will to our successors seem so simple that they will wonder why we did not find them out.

I have mentioned the external fringes seen on the illuminated side of the geometrical projection of a straight edge bounding an opaque body exposed to light coming from a luminous point, which is taken as the point of projection. This phenomenon is modified in a great variety of ways according to the outline of the opaque body. It may for instance be a screen containing one or more apertures. Suppose the aperture circular. If the circle be moderately large, and the light be received on a screen beyond, we have a circular illuminated patch corresponding nearly with the projection of the hole, which is seen to be fringed within its boundary by fringes resembling the external fringes spoken of just now. If the size of the hole be diminished, or the distance of the receiving screen increased, which produces a similar effect, the fringes invade what had previously been the uniformly illuminated area corresponding to the projection of those portions of the hole which are at a little distance from its edge ; and after a very curious and complicated set of changes we are ultimately, when the hole is made very small, left with a circular patch of rather weak light on the screen, surrounded by a dark ring, followed by other rings alternately bright and dark, but of rapidly decreasing intensity, till they are lost in the dark shadow of the screen in which the hole is pierced. The central bright patch is much larger than the geometrical projection of the hole.

The continuity of the phenomenon connects the diffusion of the light which passes through the hole in the last case, and the comparative absence of diffusion in the first case, with the formation of coloured fringes and alternations of intensity about the boundary of shadows in general, and with the external fringes belonging to a straight edge in particular, and makes it probable that if we could explain these last the principles of the explanation would enable us also to explain the existence of rays, at least in so far as rays have a real existence at all.

Can we then explain the external fringes in the simple case of a straight edge ? Newton made careful observations of the phenomenon ; but guided as he was by ideas belonging to the corpuscular theory of light, he advanced no further towards an explanation than a few vague conjectures. Dr. Young, guided by the theory of undulations, was more successful. He attributed these fringes to the interference of two portions of light, one coming direct from the luminous point, and one reflected at a grazing incidence at the edge of the opaque screen, losing half an undulation at the reflection. This theory explained very well the leading features of the fringes, showing them to be hyperbolic in form, that is to say that a section of the fringes, conceived as existing in space, made by a plane passing through the luminous point and perpendicular to the edge, is a system of hyperbolas having for their common transverse axis the line joining the luminous point with the edge, and having small conjugate axes, differing from one fringe to another of the system. It explained also the decreasing width of the fringes as we recede from the shadow, and their dilatation when the luminous point approaches nearer to the opaque screen from the edge of which they start. It even gives very nearly the breadths of the fringes and their distances from the geometrical shadow.

Accordingly when Fresnel, many years later, commenced his celebrated researches on diffraction, he in the first instance adopted Young's theory as to the cause of the formation of the external fringes. In the course however of his study he met with phenomena of diffraction which did not fall in with Young's view, and which at last opened his eyes to perceive the

grand principle which underlies the whole. We have
seen that Huygens successfully explained the laws of
reflection and refraction on the undulatory theory by
introducing the principle that each element of the
front of a wave may be regarded as the source of an
elementary disturbance, and these disturbances must
then be joined together. Now we have only to combine that principle with this other principle, that in
so compounding them we must take due account
of their respective phases, in order to account
for the whole of the phenomena of diffraction,
curious and complicated as they are. In other words,
we have only to combine Huygens's principle with the
principle of interference. These two principles again
are nothing more than special applications of the
general dynamical principle of the superposition of
small motions; a principle which lies at the very
basis of the theory of undulations, and of which the
special applications just mentioned might have been
foreseen.

The application of this principle to special cases,
among others to the case of the external fringes,
involves calculations of considerable complexity.
Fresnel executed these calculations for the external
fringes, and also made a series of most careful and
accurate measurements of the positions of the fringes
referred to the geometrical shadow under a variety of

circumstances. The theoretical distances of the several fringes from the geometrical shadow were a matter of pure prediction ; for the only unknown quantity involved in the theoretical expression, the length of a wave, had been determined by Fresnel by independent methods, some of them, as for example that depending on the measurement of the fringes produced by interference mirrors, not involving diffraction at all, so that not a single arbitrary constant was left, to be determined by some one measurement of a fringe in some one particular case, whereby an at least partial accordance between theory and observation might have been brought about. On the other hand the distances of the fringes from the geometrical shadow in a variety of cases were most carefully measured micrometrically, and the comparison of the calculated and observed places manifested a truly wonderful accordance, the average error being only about the $\frac{1}{2000}$th part of an inch.

The distances calculated from the imperfect theory of Dr. Young agreed nearly, but not exactly, with those deduced from the complete theory of Fresnel : in spite of the smallness of the difference, the measurements were sufficient to discriminate between the two, and the result was decisively in favour of the complete theory as given by Fresnel.

As the geometrical shadow is not, like the fringes,

a visible object, it may not be superfluous to mention
briefly the mode of referring the places of the fringes
to the geometrical shadow.   The fringes were formed
by the cheeks of an aperture with parallel edges, one
of which was moveable by a micrometer screw, by
which means the breadth of the aperture could be
very accurately measured.   The cheeks were set a
sufficient distance apart to prevent the fringes formed
by the one affecting in any sensible way those formed
by the other.   We thence get from similar triangles,
by the rule of three, the distance apart of the geometrical shadows at the focal plane of the eye-lens.
Now the distance of any particular fringe, say the
first dark fringe, on the right from the same fringe on
the left can be measured by a micrometer moveable in
the focal plane of the eye-lens.   Half the excess of
distance between the geometrical shadows in the
focal plane over this gives the distance of the first
minimum from the geometrical shadow.

The phenomena of diffraction may be varied indefinitely by varying the outline of the opaque body,
or aperture or apertures pierced in an opaque screen,
as well as the two distances concerned ; and there is a
large class of interesting appearances which may be
seen by using a telescope with which a luminous point
is viewed in focus, and covering the object-glass with
a screen containing one or more apertures of any form

that may be chosen. In this case especially most curious and beautiful coloured patterns are produced, so strange and complicated that a person looking at them for the first time could never guess from the pattern what was the form of the aperture which produced it.

Besides the case of the internal fringes which was so carefully examined by Fresnel, a number of other instances in both classes, that is, without and with a lens or object-glass combined with the diffracting body, have been investigated theoretically, and the results compared with observation. The accordance is found to be absolutely complete even in the most minute particulars.

And now at last, as part and parcel of the complete theory of diffraction, we are able to explain the existence of rays; to show why it is that it is so nearly true that light proceeds in a straight course past bodies and through apertures.

The explanation may be given without entering into mathematical details. Suppose light coming from a luminous point which for simplicity's sake we may suppose to be at a practically infinite distance, an element of the sun's disk for example. Let it fall on a screen in which is a moderately small aperture, and consider the disturbance produced beyond the screen at a point, *P*, well outside the projection of the aperture. Make P the centre of a set of concentric spheres with radii increasing by half the length of a wave. These will cut the plane of the aperture in a series of circular arcs, very close to one another in consequence of the extreme smallness of the wave length, and comprising between any two consecutive circles narrow slips of the aperture, such that two adjacent slips are very nearly equal in area. Now we have a right to regard each element of each slip as the source of an elementary disturbance, which reaches P after the lapse of a time proportional to its distance from P. Since corresponding points in consecutive slips differ by half a wave's length in their distance from P, and the elementary disturbances from them accordingly always reach P in opposite phases, so as to neutralize each other, we easily see that the total effect of one slip is very nearly indeed neutralized by that of its neighbour. This is still more nearly true if we take the disturbance produced by one slip and the mean of those produced by its two neighbours; and in this way, by taking each alternate slip and the mean of its two neighbours, each slip gets counted once, and once only, except at the two ends, where, however, the length of the slips dwindles away to nothing. Hence there is no sensible disturbance, and therefore no sensible light, for a point P situated as we have supposed.

If now we take a point P situated well inside the projection of the boundary of the aperture, it may be shown by similar reasoning that the disturbance, and therefore the illumination, is sensibly the same as if the screen in which the aperture is pierced were away.

This conclusion rests, it will be seen, on the extreme smallness of the length of a wave, in consequence of which an aperture, unless extremely small, is cut a great number of times by a series of concentric spheres with radii increasing by half a wave's length. There is no difference of explanation as regards light and as regards sound, save what depends on the difference of scale entailed by the difference of wave length. Take as regards light the case of a small circular hole say the tenth of an inch in diameter, and of distances from the luminous point to the screen in which the hole is pierced, and from that again to the screen on which the light is received, of say 8 feet 4 inches, or 100 inches, each. In this case, regarding the luminous patch on the screen as a whole, there would be no great diffusion of light, but the phenomena of diffraction would nevertheless be fairly pronounced. There ought to be a corresponding case of diffraction for sound; but on what scale? Take 50 inches as the length of a wave of sound, which would correspond to a musical note of moderate pitch. Taking as before the $\frac{1}{50000}$th

part of an inch as the wave length for light, the
length of the wave of sound will be two and a half
million times as great as the wave length of light.
Consequently to obtain the corresponding case of
diffraction for sound, our "small" circular hole would
be obliged to have a diameter of rather more than
4 miles, say 4 miles, and the distances from the source
of sound to the hole through which it passes, and
from that again to the place where the sound is
listened to, would have to be 4000 miles each.

It is remarkable that the existence of rays, which
formed the great stumbling-block in the way of the
early reception of the theory of undulations, is now
shown to belong to a class of phenomena, those of
diffraction, the complete and marvellously simple
explanation of which afforded by the theory of undulations now forms one of the great strongholds of that
theory.

Before leaving this subject I will briefly mention
two or three instances of diffraction which from their
paradoxical character or their importance are de-
serving of notice.

Reverting to the case in which light from a
luminous point passes through a circular aperture, and
is received on a screen beyond, consider the brightness
at a point just in the axis on the receiving screen.
If the relation between the two distances already so

often mentioned and the diameter of the hole be such
that the sum of the distances from the luminous point
to the edge of the hole and from thence to the central
point on the receiving screen exceeds the direct
distance from the luminous point to the latter by just
half a wave's length, theory shows that the illumination is actually four times as great as if the screen in
which the hole is pierced were taken away altogether.
Suppose now the hole be enlarged. We might have
said at first sight, supposing we were ignorant of the
theory, " Of course that must increase the illumination
at the central point, or at any rate cannot diminish
it." On the contrary it does diminish it; and if the
hole be enlarged till its area is just double what it
was in the first instance, the centre of the illuminated
space on the screen is a black spot. This theoretical
result is easily realized in experiment; only as the
wave length varies from colour to colour, and the
proper distance of the receiving screen varies with
the wave length, when the screen is in adjustment for
the brightest part of the spectrum it is not quite in
adjustment for the fainter ends, so that the spot
instead of being perfectly black is faintly purple.

Again, suppose a circular disk is exposed to
radiation from a luminous point, and the shadow is
received on a screen at some distance, or rather
viewed directly through an eye-lens. According to

theory, the very centre of the shadow will be a bright
point, as bright as if the disk were away. This
strange result, again, can easily be verified experimentally, easily at least if we are not too ambitious
as to the size of the disk; for the delicacy of the
experiment increases with the size of the disk, and at
the same time the total quantity of light that we
have to work with decreases. In repeating the experiment I have seen without difficulty the central
spot, with the system of rings round it well formed,
in the centre of the shadow of a disk of about the
size of a sovereign.

Among the class of diffraction phenomena in
which a luminous point or line is viewed in focus
through a telescope, and a screen with one or more
apertures is placed in front of the object-glass, there
is one case of very special interest. It is that in
which a line of light is used, suppose an extremely
narrow slit through which the sun's light is reflected
horizontally, and a fine carefully ruled grating is
placed opposite to the object-glass, the lines of the
grating being parallel to the slit. The best results
are obtained with gratings consisting of glass, or
sometimes metal, on which fine parallel lines are
ruled with a diamond point. It is requisite that the
lines should be very accurately equidistant, and in
fine gratings they are so close that several thousand

go to an inch. If the grating thus constructed be of metal, it can only of course be used for reflection.

Now on viewing through the telescope the light transmitted through or reflected from such a grating, a most remarkable appearance is presented. The luminous line is seen through or by reflection from the grating as if the ruled lines were away, and right and left of it for some way the field is dark. But then on both sides we get *pure* spectra, the blue ends being nearest to the axis. These are followed by a second set of spectra, the blue ends of which overlap the red ends of the former, and so on, the spectra as we proceed overlapping one another more and more. So pure are these spectra if the grating be a good one, that they show the fixed lines of Fraunhofer to perfection.

The formation of these pure spectra can easily be explained, and the formula expressing the deviation for light of any one kind in the spectrum of any order in terms of the interval of the grating, the angle of incidence, and the wave length for that kind of light obtained, from the general principles of the theory of undulations. But by mounting the grating in the axis of a horizontal graduated circle which carries the telescope, the deviations can be measured with extreme precision. The interval of the grating

is got by measuring the breadth of the ruled space,
and dividing by the number of ruled lines less one.
We thus have the means of determining the wave
lengths, if the theory be true, for as many definite
kinds of light marked by definite lines as we please;
and by comparing the wave lengths measured by
means of spectra of different orders, at different
inclinations of the incident light to the grating, and
by different gratings, we have a very sharp test of the
truth of the formula deduced from theory. The
accordance is complete; and that being so we are
justified in the interpretation assigned to that measurable quantity which we call a wave's length, and
obtain its value with great precision.

I may here perhaps mention that it has even been
proposed to take the length of a wave of light of some
particular kind, such for example as that belonging to
one component of the double yellow line of a soda
flame, as a natural standard to which a national
standard of length might be referred in case of loss.
The French refer their metre to the dimensions of
the earth. The English refer their yard to the
length of the seconds' pendulum. But supposing
the earth to be slowly contracting by cooling, both
these natural standards would be liable to be affected
in the course of ages; and if such a catastrophe were
to occur as the impact on the earth of some great

globe visiting our solar system, the dimensions of the earth and the value of gravity, and accordingly the length of the seconds' pendulum, would at once be affected to an unknown degree. But the wave length of light of a given kind would remain unchanged, and the survivors of such a catastrophe might have recourse to it to recover the ancient standard of length.

LECTURE III.

Closer examination of the fundamental suppositions of the Theory of Undulations—Survey of the conclusions arrived at by a study of the phenomena of common light— Elementary facts of double refraction and polarization.

IT has been my aim in these lectures to endeavour to give you some definite idea of the evidence on the strength of which we assert that light consists in undulations propagated in a medium filling the interstellar spaces. To enable you to judge more fairly of the evidence, I have attempted to present the subject in an inductive rather than in a deductive form. Instead of starting with a number of hypotheses, originating you would not know how, and then showing how the conclusions following from them are in accordance with observed results, I have commenced with only the most fundamental conceptions of undulations, and of the conditions which we must suppose to obtain in order that they may exist, and have afterwards supplemented our original rather crude conceptions in the manner which a study of the phenomena showed to be necessary. We have seen that not merely are the laws of reflection and refraction in agreement with the theory, but the curious and complicated phenomena of interference and diffraction are explained . by it to the minutest particular. What strikes one most about the theory is what has been truly described as its *astounding simplicity*. This so carries on the face of it the stamp of truth, that to one who is familiar with the phenomena as well as the theory, an overwhelming conviction is produced that it is indeed true to nature. That being the case, it may be well now to examine the various hypotheses which must be made in some greater detail than has hitherto been done.

The fundamental hypothesis of the existence of a medium to which we give the name of ether, I have already mentioned. To account for undulations in this medium, we must attach to it the two radical conceptions of inertia and elasticity. First, a finite time must be required in order to generate in a finite portion of it a finite velocity by the action of a finite force. Secondly, a change of condition of some kind in the ether consequent on a relative displacement of its parts must call into play a force of restitution tending to restore it to its primitive condition. Thus in air condensations and rarefactions produce respectively an increase and a diminution of pressure, so that any small portion of the air which has been contracted or expanded tends to push out the surrounding parts, or to be compressed by them, and so to return to its primitive state ; and it is to the forces of restitution thus called into play that the propagation of sound is due. Naturally therefore those who adopted the undulatory theory of light were led to imagine the ether as possessing a similar kind of elasticity. There is however a whole class of phenomena which I have not yet so much as named, and which have no counterpart in sound, the study of which has led us to conclude that the elasticity of the ether is of an altogether different nature from that of air. The question of the relation between the direction of vibration of the particles of ether and the direction of propagation of the light is bound up with that of the character of the elasticity by virtue of which the undulations are propagated. In the explanation, however, of the ordinary phenomena of interference and diffraction, we are not concerned with the direction of vibration ; in interference for instance, where we always have two streams of light from the same source pursuing nearly the same paths, and coming together either in the same direction, or in directions very slightly inclined to each other, whatever be the direction of vibration in one of the interfering streams, the same, or very nearly indeed

the same, will be the direction in the other, and that
is all that we are concerned with in the explanation.
And the explanation of the ordinary phenomena of
diffraction has, as we have seen, been resolved into
the interference, not of two, but of an infinite number
of infinitely small disturbances all coming originally
from the same source, and following very nearly the
same paths.

The mode of excitement of the undulations in the
first instance is in some respects analogous to the
mode of excitement of sound by a vibrating body,
such as a bell, but in some respects distinctly different.
A bell in exciting vibrations in the surrounding air
acts as a continuous elastic body ; in explaining the
mechanical action, we have nothing to do with
speculations as to the molecular constitution of
ponderable matter. But the fact that the spectra of
flames show bright lines depending on the nature of
the chemical substances in the flame, shows that in
the excitement of the ethereal vibrations we have in
many cases, if not in all, to do with the vibrations of
the constituent parts of the chemical molecules of
which we have reason to believe that ponderable
matter consists.

The explanation of refraction on the theory of
undulations requires us to suppose that in refracting
media, such as glass or water, the vibrations are

propagated more slowly than in what we call vacuum. In fact, theory shows that the sine of the angle of incidence must be to the sine of the angle of refraction in the ratio of the velocity of propagation in vacuo to the velocity of propagation in the medium. The question then arises, When light passes through water or air, what is the vibrating medium?

At first sight we might say, of course the water or the glass itself. But this supposition will not bear examination. We know too much of the elasticity of water and glass to allow us to explain the result in that way. Water is sometimes spoken of as incompressible, but we know that it yields slightly to a compressive force. The amount of compressibility can be measured by hydrostatic means, and from thence the velocity of propagation of sound in water can be determined by calculation. It comes out about four times as great as that of sound in air, which agrees very well with the experimental measurement of the velocity which has been made in the Lake of Geneva.

Glass, and isotropic elastic solids in general, possess two distinct kinds of elasticity, one, by which they resist compression of volume, the other, by which they resist a distortion not involving alteration of volume, but opposed by the force by which the substance resists the gliding of one part over another, a gliding which takes place freely in liquids. The two elastic constants of glass have been carefully determined, and from them can be calculated the two velocities of propagation of two kinds of disturbance which a large mass of glass would be capable of transmitting. These come out greater no doubt than, but still comparable with, the velocity of propagation of sound in air, and are therefore almost immeasurably smaller than the velocity required to account for the refraction of light on the supposition that it is the glass itself that vibrates. In fact, the velocity of light in vacuum is nearly 1,000,000 times as great as that of sound in air, and the velocity of light in glass would be equal to its velocity in air divided by the index of refraction of glass, or say about 600,000 times that of sound in air, so that it is altogether of a different order of magnitude.

Again, consider a gas or mixture of gases, such as air, in its relation to Light. Air can be rarefied till we have a near approach to an interstellar vacuum, or again can be condensed till its density becomes comparable with that of a liquid. Yet all through these changes there is a perfectly continuous change in its relations to light. It will not do therefore to say that light is propagated through air in one way, by one sort of mechanism, when the air is very rare, and by another when the air is very dense. But when air is rare, and makes a near approach to what we call vacuum, but which we must now conceive as space filled with the luminiferous ether, it is clear that it must be by the vibration of the ether that light is propagated in it. We are led therefore to conclude, from these considerations again, that when light is propagated in a gas condensed till its density becomes comparable with that of a liquid, and accordingly in liquids also, and in solids like glass, which behave with respect to light just as liquids, it is ether existing in the interstices between the molecules of the bodies, of which the vibrations constitute the light that passes through them.

We might not perhaps have been disposed in the first instance to suppose that such a solid material as glass really had ether pervading it. But we must beware of applying to the mysterious ether the gross notions which we get from the study of ponderable matter. The ether is a substance, if substance it may be called, respecting the very existence of which our senses give us no direct information : it is only through the intellect, by studying the phenomena which nature presents to us, and finding with what admirable simplicity those of light are explained by the supposition of the existence of an ether, that we become convinced that there is such a thing. We know that a magnet attracts iron through a piece of glass ; and yet the magnetic influence is one which we can neither see nor feel : why then should not ether exist within glass, and be capable of vibrating within it?

It may readily be imagined, as more probable than the contrary, that the presence of the ponderable molecules interspersed through the ether, within the region of space which is enclosed by the surface of the glass, may have the effect of altering the velocity of propagation of the ethereal vibrations within it, and very probably diminish it. But what may be the precise mechanism by which this result is brought about we do not know. It is easy to frame plausible hypotheses which would account for the result, but it is quite another matter to establish a theory which will admit of, and which will sustain, cross-questioning in such a variety of ways that we become convinced of its truth.

It follows from Huygens's explanation of the law of refraction, which assumes nothing but what lies at the very foundation of the theory of undulations, that the ratio of the velocity of propagation of light in vacuo to its velocity in a medium, such as glass or water, must be equal to the index of refraction, and must therefore vary from one colour to another, increasing from the red to the violet, sometimes, as in the case of oil of cassia, as much as 6 per cent. and upwards within the limits of the visible spectrum. Now we know that in sound notes of different pitch are propa-

gated with the same velocity, as also follows from
theory, and it has been suggested that it is a difficulty
in the way of the theory of undulations that the case
must be so différent with light. I cannot say that it
appears as such to my own mind ; for the case of the
vibration of one of two mutually penetrating media, as
for example glass and the ether, is so different from
any that we have to deal with in the case of sound
that we cannot argue from the one phenomenon to the
other. If indeed it were established that the velocity
of light in a vacuum differs from one colour to another,
we should then have to allow that an analogy which
might have been expected to hold good between the
two phenomena does not really do so. But till very
lately the observed phenomena which reveal a finite
velocity of light, and the experiments which had been
made to determine the velocity directly, all yielded a
negative result as to any difference of velocity between
one colour and another, so that the difference in the
ratio above-mentioned was attributable to a difference
of velocity in the refracting medium. I have said
" till lately," because in a paper by the late Dr James
Young and Professor George Forbes, presented to the
Royal Society between two and three years ago, and
printed in the Philosophical Transactions, in which
the authors have determined experimentally the velocity of light by a method founded on that of Fizeau,

but presenting certain new features, it is mentioned as a result of the observations that blue light seemed to travel faster in air than red light. The method employed is founded on the judgment of the observer as to the equality of intensity of two points of light, or artificial stars, seen simultaneously in the field of view. It is difficult to see any possible source of instrumental error which could have led to the above conclusion as to a variation of velocity depending on the colour : on the other hand, the conclusion rests on the judgment of the eye of a single observer ; and till the question has been further tested it would seem to be premature to regard the difference as established. Should it be confirmed by further observation, it will then be necessary to consider whether the circumstance that the light was propagated in a field of magnetic force, namely the earth's magnetic force, may have had something to do with it. However even if the result of further enquiry should be to show that the different colours are propagated at different rates in a simple vacuum free from disturbing influences, all we should be entitled to say is that the expectation which we were led to form beforehand from the analogy of sound, and from the supposition that the forces whereby one portion of ether acts directly on another are insensible at finite distances, at distances even comparable with the $\frac{1}{50000}$th part of an inch, has not been verified.

As regards the mode of perception, while there are analogies between sound and light there are at the same time notable differences. In sound, the tympanum of the ear is thrown mechanically into vibration, and the nerves of hearing are mechanically affected, as a mechanical disturbance of a point on the surface of the body is made known by the sense of touch. But in light, just as we have seen reason to believe that it is the disturbance of the ultimate molecules, or of their constituent parts, by which the vibratory motion which constitutes light is in the first instance communicated from ponderable matter to the ether, so we have reason to think that when light is absorbed what takes place is that the disturbance of the ether is communicated, not to portions of matter regarded as forming portions of a continuous elastic body, but to the ultimate molecules of which matter consists, or to their constituent parts. It may be that temporary chemical changes are thereby produced in the ultimate filaments of the nerves of the retina, in which case the sense of sight would be more analogous to the sense of taste than to that of touch.

Corresponding theoretically to this difference is the fact that in light we have absolutely nothing answering to the sensation of harmony in sound. When two musical notes have their times of vibration in some simple ratio, we have concord when they are

sounded together; thus the ratio of 2 to 3 gives
a fifth, that of 4 to 5 a major third &c., whereas
when two notes sounded together have their times of
vibration in no simple ratio, we have discord. But
when two kinds of light, each of definite refrangibility,
and accordingly of definite periodic time, are mixed
together, there is no pleasing or disagreeable sensation
depending on whether the periodic times are in some
simple proportion or not.

And now before passing on to a totally different
branch of the subject, the study of which leads us to
believe that the properties of this mysterious ether
must be very different from what we should have
imagined beforehand, it may be well to pause and
contemplate for a little the wonders with which our
study of the phenomena up to the present point has
shown that we are surrounded.

First, we learn to regard the interplanetary and
interstellar spaces as no mere void, or empty space
passed through by swift messengers in the shape of
particles of light conveying information from distant
worlds, but as filled with an ever present, all pervading
substance, in which the ultimate particles of ponderable
matter, including those of our own bodies, are con-
tinually as it were bathed, and yet of which our senses
give us no direct cognizance.

Secondly, that whatever other important offices

this ether may fulfil, this one at any rate belongs to it,
that it forms the medium of visual communication
between ourselves and our fellow creatures, between
ourselves and the various objects around us, between
ourselves and distant worlds.

Thirdly, that this communication is carried on by
tremors of some kind propagated through the ether
with a velocity so enormous that for all practical
purposes of communication on earth it may be deemed
instantaneous. In fact, light would travel about
seven and a half times round the whole earth in one
second. But so rapid are these tremors that many
hundreds of millions of millions take place in one second.
Notwithstanding therefore the enormous rate of propagation, the lengths of the waves are excessively
small, ranging about the $\frac{1}{50000}$th part of an inch.

It has been shown that it is this excessive smallness of wave length which enables light to be propagated so nearly in a straight course, in independent
rays. Were it not for this, the formation of sharp
images would be impossible. Were the lengths of
the waves of light comparable with the lengths of the
waves of sound, we should as regards the use of our
eyes be nearly in the condition of a man who was all
but blind; who could just distinguish light from
darkness, or a gleam of red from a gleam of green,
and no more. We should be in this condition if the

time of a vibration were anything like so great as the one hundred millionth part of a second, in which case the length of a wave of light would be comparable with that of a wave of sound.

Fourthly, we learn that notwithstanding the almost inconceivable shortness of the time of vibration, a variation in this periodic time is nevertheless recognisable by our senses, and that it is to this cause it is due that the face of nature does not present to us simply light and shade, like a photograph, but that we have that endless variety of colour which we enjoy.

Fifthly, in the plan of an elastic medium conveying small vibrations, we have a mechanism of the simplest possible kind having for result that rays of light from objects all around cross each other's paths in all sorts of ways without any mutual disturbance.

When we survey a varied landscape, each visible point in it, however minute, may be regarded as an independent source of light, from which the light proceeds in all directions. True, the objects are not in general self-luminous; they are seen by the light of the sun or of the clouds which they irregularly reflect, but as regards the behaviour of the pencils which proceed from them they are as good as self-luminous. Well then; from each visible point, however minute, there enters the eye every second a length of light of about 186,000 miles, that is, light which would have travelled that distance had not the eye been there to catch it, this immense length being filled with undulations of lengths ranging about 50,000 to the inch. And if the landscape be contemplated by a multitude of persons, from each visible point in it that vast length of light, consisting of undulations of such excessive minuteness, enters the eye of each spectator every second of time; and all these various streams of light, proceeding in all sorts of directions, cross each other's paths in all sorts of ways without the slightest mutual disturbance.

To one previously unacquainted with the subject, these statements seem like the dreams of an enthusiast, or at best the speculations of some wild theorist, and yet there is nothing in what I have stated beyond the sober conclusions of scientific investigation, conclusions supported by an amount of evidence altogether overwhelming. In saying this it is to be remembered that the precise mode of disturbance of the ether has been left an open question.

In studying this subject, one can hardly fail to be struck with the combination of these two things:— the importance of the ends, the simplicity of the means. When I say the importance of the ends, I use a form of expression which is commonly employed as expressing design. And yet on that very account we must be on our guard against too narrow a view. When we consider the subject of vision in its entirety, the construction of the recipient organ as well as the properties of the external agent which affects it, the evidence of design is such, it seems to me, as must to most minds be irresistible. Yet if I may judge of other men's minds by my own, it is rather in the construction of the recipient organ than in the pro- perties of the agent that affects it, that the evidence of design is so strongly perceived. And the reason of this may be that we are here dealing with what more nearly resembles design as we know it in our- selves. Man takes the laws of matter as he finds them; the laws of cohesion, of the conversion of liquid into vapour, of the elasticity of gases and vapours, and so forth; and in subserviency to those laws he constructs a machine, a steam-engine for instance, or whatever it may be; but over the laws themselves he has absolutely no control. Now when we contemplate the structure of the eye, we think of it as an organ performing its functions in subserviency to laws definitely laid down, relating to the agent that acts upon it, laws which are not to be interfered with. We can it is true go but a little way towards explaining how it is that through the intervention of the eye the external agent acts upon the mind. Still, there are *some* steps of the process which we *are* able to follow, and these are sufficient to impress us strongly with the idea of design. The eye is a highly specialized organ, admirably adapted for the important function which it fulfils, but, so far as we can see, of no other use; and this very specialization tends to make the evidence of design simpler and more apparent. But when we come to the properties of the external agent which affects the eye, we begin to get out of our depth. These more nearly resemble those ultimate laws of matter over which man has no control; and to say that they were designed for certain important objects which we perceive to be accomplished in subserviency to them, seems to savour of presumption. It is but a limited insight that we can get into the system of nature; and to take the very case of the luminiferous ether, while as its name implies it is all important as regards vision, the present state of science enables us to say that it serves for one object of still more vital importance; we seem to touch upon another; and there may be others again of which we have no idea.

In the study of those phenomena of light which I have hitherto brought before you, we derived constant assistance from our knowledge of the theory of sound.. I now come to a branch of the subject where the theory of sound fails us altogether, to a class of optical phenomena which have nothing answering to them in sound. I refer to double refraction and polarization.

Of these phenomena, the former of which is so closely related to the latter, double refraction was the first to be discovered. It was in 1669 that Bartholinus published his account of the discovery of a strange and unusual refraction in Iceland spar. The subject was taken up and investigated with the keenest interest by Huygens, whom we must regard as the founder of the theory of undulations. It would not be in accordance with the object which I have had in view in these lectures to enter into details respecting the phenomenon, and I must content myself with mentioning a few of the more salient features.

Iceland spar stands we may say alone among minerals in at the same time possessing powerful double refraction and occurring in large clear crystalline masses. It was this circumstance which led to the discovery. The mineral cleaves very readily in three definite directions, so that a block obtained by cleavage is of the form of a parallelepiped. The three obtuse dihedral angles of this parallelepiped are all exactly equal, and are so turned that two opposite solid angles are contained by three equal obtuse plane angles, while each of the remaining six is contained by one obtuse and two acute. A *direction*

—not, observe, any special line—equally inclined to
the three edges which meet in one of the obtuse solid
angles is called the *axis* of the crystal. The crystalline structure, so far at least as it is revealed by the
cleavage planes, or I may add by the natural faces, is
symmetrical with respect to three planes and no
more, each of these being parallel to one of the edges
which meet in an obtuse solid angle, and perpendicular to the plane of the other two, and accordingly
being parallel to the axis of the crystal. Any one
of these planes is what Huygens called a principal
plane.

If an object near at hand be viewed through such
a block, as for example if the block be laid on a
printed page, two images of the object are seen, of
which one appears more elevated than the other when
both eyes are used, indicating a stronger refraction.
This image is called the ordinary, because it obeys
the ordinary law of refraction. That such is the
case was found by Bartholinus, and confirmed by
Huygens; and the accuracy of the ordinary law as
applicable to this pencil had stood the test of the
most refined measurements carried out by the most
improved modern methods, among which I may
specially mention the measurements made by Professor Swan and Mr Glazebrook. But the rays
belonging to the other or *extraordinary*, as it is called,

image must obey some totally different law. If the eye and object be in the principal plane of the block, the extraordinary ray obeys the laws of ordinary refraction so far as this, that the refracted ray lies in the plane of incidence; but whereas in ordinary media there is no refraction at a perpendicular incidence, and at oblique incidences there *is* refraction, which goes on increasing as the angle of incidence increases, in the case we are now considering there *is* refraction at a perpendicular incidence, and at one particular oblique incidence a ray passes through *without* refraction. And if the block be turned round in the plane of its surface, so that the plane through the eye and the object, which I suppose perpendicular to the plane of the surface, is no longer a principal plane of the crystal, the refracted extraordinary ray does not so much as lie in the plane of incidence.

If a distant object be viewed through the block, there is no duplication, so that whether a ray passes through the block as ordinary or as extraordinary, the emergent ray to which it gives rise is parallel to the incident, just as in the case of ordinary refraction. This is true whatever be the inclination of the incident ray to the surface, and whatever be the plane of incidence. It depends however on the parallelism of the surfaces of incidence and emergence; and if these surfaces be inclined to each other, forming a prism,

then the duplication of a distant object is at once
perceived. If a slit of light be viewed through such
a prism, two spectra are seen, which are unequally
deviated, and show in general unequal dispersion.

Now what notion are we to form of the cause of
double refraction if we adopt the theory of undulations? According to the fundamental explanation
of refraction given by Huygens, as there are two
refracted rays, there must be two disturbances propagated within the crystal as the result of an elementary disturbance excited at a point of its surface, and
these two must travel with different velocities. They
will spread out from the centre of disturbance in two
closed surfaces respectively, or it may be a surface
with two sheets. For one of these, the velocity of
propagation will be the same in all directions, and the
surface will therefore be a sphere. This follows from
the fact that the ordinary ray obeys the ordinary law
of refraction. But as the other obeys some more
complicated law, the surface for it must be other than
a sphere, and the velocity of propagation must be
different in different directions.

It has been already remarked that there are three
planes of crystalline symmetry, and these, as might
have been expected from the intimate relation of the
optical properties to the crystalline structure, are
also planes of optical symmetry; that is to say, all the

optical properties are symmetrical with respect to each of them. But there are other planes which are planes of optical, though not of crystalline, symmetry. Thus if a plate be cut perpendicular to the axis, the optical properties in it are symmetrical with respect to any plane through the axis; and if a plate be cut parallel to the axis, the plane perpendicular to the axis is a plane of optical symmetry as well as the plane passing through the axis, though the former cannot be, and the second is not in general, a plane of crystalline symmetry. In short, all the optical properties are symmetrical about the axis, the two poles of which are alike, so that the optical properties present the same degree of symmetry as an ellipsoid of revolution*. Accordingly the wave surface relating to the extraordinary ray must have thus much symmetry; and Huygens assumed for trial that it was a spheroid of revolution. As far as he could make out, the refraction in the direction of the axis appeared to be the same for the two rays, and he accordingly supposed that the sphere belonging to the ordinary ray, and the spheroid which he assumed

---

* According to Sir David Brewster, this is not altogether the case so far as relates to the properties of the light reflected from an artificial surface of the crystal. His observations appear never to have been published in detail, nor has anyone else, so far as I know, taken up the subject.

as the form for the extraordinary, touched one another in the axis. This equality of refraction along the axis we now know to be rigorously exact. The form of the wave surface being assumed, the refraction of the extraordinary ray followed at once from Huygens's construction; and the mode and amount of refraction were found to agree with the construction as near as the most accurate measurements made by Huygens could decide. Mr Glazebrook has recently executed a series of measurements of the refraction of the extraordinary ray in Iceland spar with all the exactitude of modern methods, guided by our increased knowledge of the subject; and the result is that no certain error of Huygens's construction can be detected.

So far however the laws of the extraordinary refraction in Iceland spar are merely empirical, based upon a happy guess as to the form of the extraordinary wave surface; it remains to be explained, if explain it we can, why there should be these two kinds of disturbance at all within the crystal, and why the form of the wave surface should be what we find it must be if the most fundamental principles of the wave theory are true, by which the form of the surface is connected with the observed refraction.

Huygens imagined that the ordinary ray was propagated by the vibration of the ether within the crystal, while in the extraordinary the molecules of the crystal took part as well as the ether. I have already mentioned some of the strong objections which exist to the supposition that the propagation of light in media such as water or glass takes place by vibrations of the ponderable matter; and similar objections in good measure apply to the supposition that both the molecules and the ether take part in it. But be that as it may, Huygens himself, when he had nearly concluded his researches, discovered a remark- able phenomenon which ill accords with the supposi- tion of his which I have just mentioned.

Suppose a second block of Iceland spar to be placed on top of the first, so that the two are in the same relative position which they would occupy if they formed one larger block. Then neither of the images, ordinary or extraordinary, seen through the first block is split into two in passing through the second block, but the ordinary of the first furnishes an ordinary in the second, and nothing more, and simi- larly the extraordinary an extraordinary. The same is still the case if the second block be separated from the first, or even inclined by turning in the principal plane. But reverting for simplicity's sake to the first relative position of the blocks, suppose that the second is turned round the common normal to the two adjacent surfaces. The moment the second block is

S. 7

turned from the primitive position, each of the images
which the first block furnishes is split into two by
the second block. The relative positions of the two
images forming the pair into which either of the
original images is split is just the same as if that
original image had been formed by common light,
but the intensities are different. When the second
block has been turned through only a small angle,
the ordinary of the first furnishes mainly an ordinary
in the second, but also a faint extraordinary, and
similarly the extraordinary of the first furnishes mainly an extraordinary in the second, but also a faint
ordinary. As the turning goes on, the faint images
get brighter, and the bright images get weaker, till
after turning through 45° the four are alike. As the
rotation continues, the pair of images that had been
the weaker get the brighter, and the pair that had
been the brighter get the fainter, till after turning
through 90° the pair of images that had at first
appropriated the whole of the light disappear altogether, and the pair that began to spring into existence on first turning now alone are seen. In this
position, that is to say when the principal planes of
the two blocks are perpendicular to each other, the
image formed by light which suffered ordinary refraction in the first block furnishes nothing but an extraordinary in the second, and the image formed by

light which suffered extraordinary refraction in the first furnishes nothing but an ordinary in the second. As the rotation goes on, the same series of changes are repeated, so that in a complete revolution the two pairs of images vanish alternately at every quarter of a revolution.

On account of the fundamental importance of this phenomenon, I must crave your indulgence for dwelling on it at some length, even though from its elementary character it must be familiar to those of you who have paid any attention to this branch of the subject.

Suppose that light coming directly from a luminous source, such as the flame of a lamp, is received on a screen with a circular hole. The screen will isolate a beam of light which, as I shall have occasion to deal with it only at short distances from the hole, I will call cylindrical. Let this beam be received, suppose perpendicularly, upon a block of Iceland spar bounded as usual by cleavage planes. In passing through the spar, it will be divided into two, an ordinary beam, which will pass straight on, and an extraordinary which will be deviated in a lateral direction in passing through the block, and will give rise to an emergent beam parallel to the incident, and accordingly parallel to the first emergent beam. If the hole in the screen be not too large to suit the thickness of the block, the two beams will come out without overlapping, and may be examined apart. They are found to be of sensibly equal intensity whatever be the azimuth of the block around its normal. Making abstraction of the small quantity of light which goes elsewhere by reflection, we may say that the whole of the incident light is divided equally between the two beams.

Suppose now that we fix the block in any position, say with its principal plane vertical, and place a second screen with a circular hole to let pass the beam which went straight through the first block, while it stops the other. On examining this beam by a second block, which is turned round, we find that it is divided into two of unequal intensity, which vanish alternately at every quarter of a revolution, the whole of the light passing into an ordinary beam in the second rhomb when its principal plane is vertical, and into an extraordinary when it is horizontal. Hence whereas a beam of common light is propagated in some definite direction, but possesses no relation to space in any other direction, so that there is no plane passing through it which we can distinguish, merely by the properties of the beam itself, from any other, the beam we are now considering, namely that which passed through the second screen, possesses properties with respect to directions in space transverse to its direction of propagation ; and if we knew nothing of its history, but it were merely presented to us for examination, that would not hinder us from recognising, by means suppose of a rhomb of Iceland spar, the peculiar properties which it possesses, nor from fixing by observation alone the direction of those two rectangular planes, vertical and horizontal in the case supposed, with respect to which its properties are symmetrical, and in either of which if the principal plane of the examining rhomb be placed, an ordinary or an extraordinary beam, as the case may be, is alone produced in it. Light possessing this property, however it may have acquired it, is said to be polarized, and the plane with respect to which its properties are the same as are those of the ordinary ray in Iceland spar with respect to the principal plane, is called the plane of polarization. Its azimuth may be determined experimentally as being that of the principal plane of an examining rhomb when so turned as to transmit only an ordinary beam.

To go back now to the first arrangement, namely, that of a beam isolated by a first screen falling on a rhomb of Iceland spar, and then on a second screen provided like the first with a suitable hole, let this second screen be so placed as to transmit the beam which passed through the rhomb by extraordinary refraction, and stop the other, and then let the beam be presented for observation. On examining it with a

second rhomb we should find that it possessed identically the same properties as the beam obtained by
ordinary refraction in the first rhomb, save that the properties of this beam are related to the horizontal plane
precisely as were those of the former beam to the vertical plane ; and it is into the horizontal plane that the
principal plane of the examining rhomb must be turned
in order that nothing but an ordinary beam may be
transmitted through it. Hence, according to our
former definition, we must say that the extraordinary
beam passing through Iceland spar is polarized in a
plane perpendicular to the plane of incidence.

More than a century elapsed after Huygens's
discovery of what we now call polarization before
it was known that polarized Light could be obtained
otherwise than by or as an accompaniment of double
refraction. But in 1808 Malus made the very important discovery that when Light is reflected from glass
at a certain angle, the reflected ray is wholly polarized ;
and since the properties of the reflected ray are the
same with reference to the plane of incidence as are
those of the ordinary ray in Iceland spar with reference
to the principal plane of the crystal, we must in
accordance with our definition say that the reflected
light is polarized in the plane of incidence. The
light reflected at other inclinations possesses all the
properties of a mixture of common light with light

polarized in the plane of incidence, and may accordingly be said to be partially polarized in the plane of
incidence. The transmitted light, whether the light
be incident at the polarizing angle or the angle of
incidence be arbitrary, is found to be partially polarized in a plane perpendicular to the plane of
incidence.

Malus found that Light is thus polarized by
reflection from transparent substances in general,
from varnishes &c., but not from metals, the light
reflected from which is found to be only partially
polarized in the plane of incidence. The angle of
incidence on transparent substances required for complete polarization Malus found to vary from one to
another, though he did not discover according to
what law, a law afterwards made out by Brewster,
namely, that the polarizing angle is that for which
the reflected and refracted rays are perpendicular to
each other ; or, as it may be otherwise expressed, the
tangent of the polarizing angle is equal to the index
of refraction.

Malus's important discovery of the polarization of
Light by reflection proves that polarization, whatever
it may be, is something that may exist altogether
independently of double refraction, and must therefore
be something intimately bound up with the nature of
Light in itself. The intimate connexion of double

refraction with polarization shows that we cannot hope to explain the former of these phenomena unless we can obtain some insight into the nature of what constitutes polarization.

LECTURE IV.

Phenomena presented on interposing a crystalline block or thin plate in the path of Polarized Light which is subsequently analyzed—Laws of Interference of Polarized Light— Theory of Transverse Vibrations—Conclusion.

WE have seen that when a beam of polarized light is examined by a rhomb of Iceland spar, it is divided into two of in general unequal intensity passing through the spar. As before, make abstraction of the small quantity of light which goes elsewhere by reflection, and call the intensity of the incident beam unity, and let us consider in the first instance the intensity of the beam which passes through the examining spar as ordinary. Let A be the azimuth of the principal plane of the examining rhomb referred to the plane of primitive polarization. Then the intensity, the relation of which to the angle A is the object of our search, must be such that it is equal to unity when A is nothing, decreases to nothing as A increases to $90°$, increases to unity again as A increases to $180°$, decreases again to nothing as A increases to

270°, and finally increases to unity as at first as A increases to 360°, having furthermore the same value for A negative as for A positive, and for $90° - A$ negative as for $90° - A$ positive. A very simple function possessing this property is $\cos^2 A$. If this be the intensity of the ordinary, since the rest of the light passes into the extraordinary, the intensity of the latter must be $1 - \cos^2 A$, or $\sin^2 A$, or the squared cosine of the angle between *its* plane of polarization and the plane of primitive polarization. Such was the law assumed by Malus, and called after his name. It has been verified by photometric measurements, and is of great importance with reference to the theory of what it is which constitutes polarization.

Suppose that light polarized in any way is subsequently analyzed, as it is called, whether by a thick block of Iceland spar furnished with screens so as to stop out one of the transmitted pencils, or by any of the more convenient methods more commonly employed. Let the analyzer be turned till the field of view is dark, the light falling upon the analyzer in that position being stifled, as in the case of a good tourmaline, or else sent elsewhere. If a block of Iceland spar be interposed between the polarizer and the analyzer, and turned round in its own plane, in general there is more or less restoration of light, there being only four azimuths of the block, separated by 90°, in a complete turn in which the field remains dark as before the interposition of the block.

This restoration of light is very easily explained as a consequence of what has already been mentioned. The polarized light falling on the block is divided into two beams polarized respectively in and perpendicularly to its principal plane, which traverse the crystal independently though overlapping, and which emerge blended together. Each of these on entering the analyzer is again divided into two, polarized in rectangular planes, which are those of the plane of primitive polarization and the perpendicular plane, of which the latter alone is retained, and the two retained portions of each of the streams enter the eye together, and their illuminations are added together. A very simple application of Malus's law shows that if we take the intensity of the primitively polarized light for unity, and disregard the small loss by reflection, the intensity of the light entering the eye will be half the square of the sine of double the azimuth of the block, measured from one of the vanishing positions. This restoration of light in the dark field forms a very sensitive and easily applied test of the possession of double refraction by the substance interposed.

But an observation made by Arago about 1811 opened out quite a new field of research, remarkable alike for the beauty of the phenomena, for the light they throw upon the nature of polarization, and for the information they afford us respecting the inmost structure of bodies. Arago found that when the interposed crystalline plate was very thin, as may easily be obtained with mica or sulphate of lime, the restored light was not as before white, but showed the most gorgeous colours, varying with the thickness of the plate, its nature, and the direction in which and amount by which it is inclined, if inclined it be.

The more powerfully doubly refracting be the substance interposed, the thinner as a rule is the interposed plate required to be in order to show these colours. But with a given substance, such for example as Iceland spar, the amount of double refraction varies immensely with the direction. Thus in Iceland spar, one of the most powerfully doubly refracting substances known, we have seen that in the direction of the axis the two rays are refracted alike. Accordingly if a plate of Iceland spar, even a thick plate, be cut perpendicular to the axis, and be interposed perpendicularly to the incident light in the dark field, a splendid system of coloured rings is seen, which are interrupted by a dark cross. The arrangement of coloured curves is still more remarkable in the case of biaxal crystals, but the simpler case of a thin crystalline plate is

better adapted to our present purpose, since in the
other case there are too many things crowded at
once upon the attention.

Let us revert then to the case of a thin crystalline
.plate interposed in the dark field, being held, as
I will suppose, perpendicularly to the incident light.
Even with this restriction, it would take too long to
mention all the phenomena exhibited, and would be
wearisome; nor is it advisable to treat the subject
in this way, for in fact they have all been brought
under the dominion of theory, and are best studied
in connexion with it, except in so far as may be
necessary to establish the theory in the first instance.

I shall restrict myself therefore to mentioning a
few leading features of the phenomenon, premising
that in the case of a doubly refracting plate in
general, as in that of a block of Iceland spar, there
are two rectangular directions in which the beams
independently transmitted are respectively polarized,
and that if the incident light is polarized, and the
plane of polarization coincides with either of those
directions, the whole of the light entering the crystal
passes into the beam which is polarized in the plane
of primitive polarization.   The two rectangular directions above mentioned have been named the *neutral
axes* of the plate.

If the thin crystalline plate interposed be of uniform thickness, it is seen of a uniform colour; if the thickness vary irregularly, as in the case of a plate of selenite obtained by casual cleavage, the colours are arranged in patches, corresponding to the varying thickness. If the plate be turned in its own plane, there are four rectangular positions in which the field is left dark as at first, which are those in which one or other of the neutral axes lies in the plane of primitive polarization, and in which accordingly there is no bifurcation of the incident light on passing into the crystal.

If the analyzer be turned through 90°, so that the planes of polarization of the polarizer and analyzer are now coincident instead of perpendicular, and the field before the introduction of the crystal is at its maximum of brightness, on interposing the crystalline plate the colours now seen are complementary in character to the former. They vanish altogether when either neutral axis comes into the plane of primitive polarization, and are less vivid than in the dark field except when the neutral axes are at an azimuth of 45° from the vanishing positions.

If the planes of polarization and analyzation be set at an arbitrary angle, and the crystal be turned in its own plane, there are eight positions in a complete revolution in which the colours disappear, giving place to white light of the same intensity as when the plate is away. Between these critical positions, the colours have the character of those of the dark and of the bright field alternately. The critical positions are those in which one of the neutral axes lies in the plane of polarization or analyzation. Hence— and this is to be specially noticed—for the production of the colours it is essential that the polarized light we start with should be divided into two pencils in passing through the crystal, and that each of these again should be divided into two by the analyzer, of which one portion is retained.

If the crystalline plate be ground into the form of a slender wedge, the colours are arranged in bands parallel to the edge of the wedge, the bands for any colour being equidistant, and the scale larger for the red than for the blue. If the analyzer be set to give the dark and the bright field in succession, the tints of the wedge agree with those of the reflected system of Newton's rings in the former case, and the transmitted system in the latter.

The equidistance of the bands for any particular colour shows that the *law* of the order of the tints, as depending on the thickness of the plate, is the very same as in the case of Newton's rings, the *magnitude* of the thickness merely being very much greater than in that case. Nor is this all. If we know the doubly refracting energy of any particular

substance, suppose sulphate of lime, we can calculate
the retardation of phase of one relatively to the other
of the two rectangularly polarized pencils which a
thin plate of the substance can independently transmit, in terms of the thickness of the plate, to which
that retardation is proportional.   Now Dr Young
showed that when this is done the thickness of plate
by which any particular tint is produced is just what
it ought to be *on the supposition* that the colour is
due to the interference of the two rectangularly
polarized pencils which traversed the crystal independently.   After all this we can hardly help supposing that the colours must in some way be due
to interference.   But if so, why are they not seen
with common light, just as Newton's rings ; why
should it be necessary that the light should in some
way be polarized, and the polarized light should in
some way be analyzed, and that the crystalline plate
should be interposed *between* the polarizer and the
analyzer in order that any colours at all should be
seen ?

  If we *assume* that the colours of crystalline plates
in polarized light are due to interference, the laws
of the interference of polarized light may be deduced
from the observation of those colours without any
experimental difficulty.   But if it still be regarded as
in any way an open question whether those colours

are due to interference, it becomes important to investigate those laws by means of experiments free from any such doubt.

Accordingly the series of researches by which Arago and Fresnel determined in a direct manner the laws of interference of polarized light must be regarded as making an epoch in the progress of the study of this branch of the subject. These experiments were made on the fringes of interference with which we had already become familiar as exhibited by common light; such fringes as those produced by the interference of two streams of light from the same source, such as two virtual images of a luminous point, or of two streams from the same luminous point, which after passing through two parallel extremely narrow apertures near one another diverged and mixed together. We have seen what a triumphant explanation the theory of undulations affords of the phenomena of interference and diffraction in the case of common light; and if we obtain the same fringes, with the appearance of which we are so familiar, in working with polarized instead of common light, we cannot refuse to admit that they too are due to interference.

The study of the interference of two streams of light polarized in the same way presents no experimental difficulty whatsoever. It is merely necessary

S.   8

to use polarized light instead of common light in
any of the ordinary experiments of interference.   On
doing this the phenomena of interference are found
to be absolutely the same with polarized as with
common light.   But to polarize two portions of light
from the same source, and proceeding nearly along
the same course, in two rectangular directions, and
yet ensure a very near equality in the lengths of their
paths, or rather equivalent paths in air, is a matter
of very great nicety, so small is the difference of
path that would suffice to prevent any exhibition
of interference of the usual kind.   Nevertheless by
a series of ingeniously devised and carefully executed
experiments Arago and Fresnel succeeded in establishing conclusively under what circumstances polarized light is, and under what it is not, capable of
manifesting the usual phenomena of interference.

The result of this enquiry was summed up in five
laws relating to the interference of polarized light,
which were derived directly from observation.   One
of these has already been mentioned.   Another is
that when two streams of light from the same source
are polarized in rectangular planes, they show no
phenomena of interference, notwithstanding a near
equality of paths.   Another, that the mixed stream
as in the last case may be analyzed without any
phenomena of interference being thereby revealed.

Another that when two streams of light from the same source are polarized in rectangular planes, and afterwards analyzed, they *do* manifest the phenomena of interference *provided* the original source were one of polarized instead of common light. Lastly, in the phenomena of interference produced by rays which have experienced double refraction, the place of the fringes is not in all cases determined solely by the difference of equivalent paths; in certain cases it is necessary to alter the difference of paths by half an undulation. And the rule they gave for determining under what circumstances the half undulation should be added and under what circumstances not amounted to this:—when the planes of polarization and analyzation lie in the same quadrant made by the neutral axes of the crystalline plane, the character of the interference is determined simply by the difference of paths; but when they lie in adjacent quadrants, we must alter the difference of paths by half an undulation.

These five empirical laws embrace the necessity for a polarizer and for an analyzer in order that colours should be seen in a crystalline plate; and taken in conjunction with Malus's law, regarded at present merely as an empirical law, they enable us to calculate completely the colours of crystalline plates under all the varied conditions which may exist as

to thickness of the plate, doubly refracting energy
of the substance of which it is formed, azimuth of
the neutral axes relatively to the plane of primitive
polarization, azimuth of the plane of analyzation;
and for that we have no occasion to enter into any
speculation at all as to what it is that constitutes
polarization. Nay more; if we take the laws of
double refraction as known empirically as the result
of direct observation, we may even calculate completely the coloured rings and curves about the optic
axis or axes of uniaxal or biaxal crystals, without
entering into any speculation as to what the theoretical interpretation of polarization may be.

The question now arises, can we embrace the five
laws of interference of polarized light, and Malus's
law, in a theory which shall comprehend them all,
and which shall be at least hopeful for the explanation
of double refraction and of the polarization of light by
reflection; though as these may depend on a knowledge of what is the actual state of things which we
do not possess, we cannot demand of necessity that
it *shall* lead to their explanation.

In applying Malus's law to the calculation of the
colours of crystalline plates, we are led to contemplate an intensity which we may take as unity in the
incident original polarized light as giving rise to
intensities $\cos^2 A$ and $\sin^2 A$ belonging to light

polarized in a plane making an angle A with the plane of primitive polarization and in the perpendicular plane respectively, and these again as giving rise to polarized beams the intensities of which are obtained by a re-application of the very same law. Now in studying the interference of common light, we saw reason to conclude that for light of a given kind, that is, of a given refrangibility, the intensity is measured by the square of the coefficient of vibration, and consequently the coefficient of vibration by the square root of the intensity. It is true that I have not been able to lay before you the full evidence on which this conclusion is based, as it would have involved some considerations of too mathematical a nature to be suitable to the present lectures, so that I have been obliged to leave the result to be accepted in some measure on the strength of authority. Consequently we are led by pure observation, combined with so much of theory as belongs to the study of common light, to contemplate a beam of polarized light in which the coefficient of vibration may be taken as unity as being divided (as for example in passing through a crystalline lamina) into two polarized in rectangular planes, at azimuths of A and $90^0 - A$ to the plane of primitive polarization, the coefficients of vibration belonging to which are expressed by $\cos A$ and $\sin A$. But this is

identically the law according to which forces, or displacements, or velocities in directions *perpendicular* to
that of propagation, and in or equally inclined to the
planes of polarization, would be resolved. We are
inevitably driven to the contemplation of a *something*
about polarized light which admits of composition
and resolution according to the above simple law.
This "something" can hardly be other than the
vibrations themselves, and we are thus led to conclude
that in polarized light the vibrations are rectilinear,
but instead of being in the direction of propagation,
as from the analogy of sound the vibrations of Light
might naturally have been expected to be, are *transverse* to the direction of propagation.

When polarized light is obtained by ordinary
refraction through Iceland spar, or by reflection from
glass at the proper angle or incidence, everything is
symmetrical with respect to the plane of polarization.
We must suppose therefore that in polarized light
the vibrations are symmetrical with respect to the plane
of polarization. This leaves two alternatives open :
they may either be in the plane of polarization or
perpendicular to the plane of polarization. So far as
the explanation of the laws of interference of polarized
light is concerned, it is a matter of absolute in-
difference which alternative we adopt, and some
undulationists have adopted the one and some the

other. The question can only be decided, if it can be decided at all, by introducing more or less of dynamical considerations, and that introduces more or less of speculation, since the dynamical nature of the mutual action of ponderable matter and ether is in great measure unknown to us. Perhaps the argument which introduces least speculation as to the dynamical nature of such actions is that derived from diffraction at a considerable angle; though it is true that even here we cannot produce that diffraction without the intervention of ponderable matter. The result in this case, as I have elsewhere shown, is decidedly in favour of the supposition that the vibrations are perpendicular to the plane of polarization, which is the alternative that was adopted by Fresnel, and ultimately by Cauchy, though at first he adopted the other; and it is the one for which, independently of diffraction, there is I think the most to be said. But as I have remarked the theory of transverse vibrations taken by itself does not involve the decision of this question.

If such be the nature of polarized light, what notion are we to form as to the nature of common light? Suppose light coming in some definite direction to fall on a screen with a hole, and to be received at the other side on a block of Iceland spar. Then if the hole be not too large, the two beams which are produced within the spar will come out separated from one another, and will show their polarization in rectangular planes. This will still be true however the screen may be moved about, so that the light falls on different parts of the face of the block. Now let the screen be removed altogether. The light will still be decomposed into two beams within the spar, giving rise on emergence to two beams polarized in rectangular planes, but the beams will be broad, and will mix on emergence. But the mixture is identical with, or at any rate is undistinguishable from, common light.* Now the two disturbances in rectangular planes give rise by their composition to a disturbance which is in the fronts of the waves, but is in general elliptical, including the extreme cases of circular and rectilinear vibrations, but with the elements of the ellipse changing, as we have every reason to expect, irregularly in all sorts of ways a great number of times in a second. For though the vibrations may be sensibly regular for thousands or it may be myriads together (and the phenomena of interference show that they must have

---

* Abstraction is here made of the loss by reflection, which is not quite the same for the ordinary and extraordinary, the higher refraction of the former being accompanied by a slightly more copious reflection, so that in the transmitted light there is a slight theoretical preponderance of intensity in favour of the extraordinary. This is however so small as to be barely sensible in refined experiments, and for our present purpose it is best neglected.

a high degree or regularity) yet we should have no reason *a priori* to expect that they would remain regular for the fifty millions of millions or so of vibrations which must take place in the tenth of a second, the time for which an impression made on the retina is estimated to last.

Such being our notion of common light, the division of common light into two rectangularly polarized beams which follow different paths must be taken to imply that for some cause yet to be investigated the vibrations, which at first were in the fronts of the waves, but in other respects of any kind, are decomposed into two rectilinear vibrations in rectangular directions, which are propagated along different paths.

This fundamental conception as to the nature of polarized light, and its relation to common light, explains in the simplest manner the six laws relating to the interference of polarized light, and to the intensity of the polarized streams into which polarized light may be divided, to which I have just referred. The interference of light merely demands, so far as direction of vibration is concerned, that it should be as nearly as possible the same for the two interfering streams, a condition satisfied by two streams polarized in the same way. Again, the kinetic energy of a set of vibrations, to which for light of a given kind the intensity is theoretically proportional, is the sum of

the kinetic energies of the components in any two
rectangular directions, irrespective of their difference
of phase, and therefore no phenomena of interference
ought to be visible in the mixture of two rectangularly
polarized portions of light, even though they came
originally from the same polarized source. Again, if
common light be divided into two rectangularly
polarized portions, which are afterwards subdivided
in a similar manner, and a pair of these latter components which are polarized alike mix, having had
but a small difference of path from the original source,
they ought not to show any signs of interference.
For as there is no fixed and permanent relation
between the relative magnitudes or the relative
starting-times of the first components, they are as
good as if they came from two independent sources,
in which case no phenomena of interference are
either theoretically observable or experimentally observed, do what we will with the streams of light
afterwards.   But if the original source of light be one
of polarized instead of common light, the case is
altogether different.   Then, whatever changes take
place during the tenth of a second in the amplitude
or starting-time of one of the first components, exactly
the same take place in the other, and are carried on
into the second components, of which therefore those
which are polarized in the same way are in a condition

to interfere. And as to the circumstances under which in this kind of interference the difference of path must sometimes be imagined altered by half an undulation, the matter is simple to the last degree on the theory of transverse vibrations; it merely involves attention to the sign of a geometrical projection; and if we express the intensity of the mixed light, for the case in which the planes of polarization and analyzation lie in the same quadrant formed by the neutral axes of a crystalline plate, by a formula in which the symbols denote displacements or velocities or of an ethereal particle instead of intensities, the formula takes care of itself, and applies equally to the case in which the planes of polarization and analyzation lie in adjacent quadrants.

If polarized light be incident on a crystalline plate of uniform thickness, and the emergent light be viewed through an analyzer which is turned round, then except in certain special cases which I need not particularize, the light is not extinguished by the analyzer however it be turned, but merely becomes, in general, alternately a maximum and a minimum alternately at every quarter of a turn. So far it agrees exactly with partially polarized light. And yet the two are altogether different, and the difference may be seen at a glance by viewing them through a Nicol's prism capped by a plate of Iceland

spar cut perpendicular to the axis. The light we are
now considering is called elliptically polarized light,
and in contradistinction to it the light which I have
hitherto called polarized has been denominated plane-
polarized.

Elliptically polarized light may be obtained independently altogether of double refraction. The theory
of transverse vibrations presents to the mind a very
clear picture of what constitutes it. But as this
is a matter of detail not involving any fresh principle,
I forbear to enter further into it.

Being now armed with a definite theory as to the
nature of polarized light, we are prepared to consider
whether any explanation can be given of double
refraction, as an accompaniment of which polarization
was first discovered. Whether we are able or not to
give a complete explanation of it, we might expect
that the theory would at least so far fall in with it as
to point hopefully towards an explanation.

The most salient feature of double refraction,
interpreted by the aid of the theory which makes
light to consist in undulations, and of that which
specifies the nature of polarized light, is that when
light falls on a crystalline plate or prism it gives rise
to two kinds of disturbance within the crystal, which
are propagated with different velocities, and in which
the vibrations are rectilinear, and take place in planes

which are perpendicular to each other, or at least very approximately so, and of which the directions are determined by lines fixed in the crystal.

Now we have not far to go to find a mechanical illustration of such a mode of action. Imagine an elastic rod terminated at one end, and extending indefinitely in the other direction. Let the rod be rectangular in section, the sides of the rectangle being unequal, so that the rod is stiffer to resist flexure in one of its principal planes than the other. Let this rod be joined on to a cylindrical rod forming a continuation of it which extends indefinitely. Conceive the compound rod as capable of propagating small transverse disturbances in which the axis of the rod suffers flexure. Imagine a small disturbance, suppose periodic, to be travelling in the cylindrical rod towards the junction. It will travel on without change of type even though the flexure of the axis be not in one plane. But to find what disturbance it excites in the rectangular rod, we must resolve the disturbance in the cylindrical rod into its components in the principal planes of the rectangular rod, and consider them separately. Each will give rise in the rectangular rod to a disturbance in its own plane, but the two will travel along the rod with different velocities. This illustrates the subdivision of a beam of common light falling on a block of Iceland spar into two beams polarized in rectangular planes, which are propagated in the spar with different velocities. Again, suppose the original disturbance in the cylindrical rod confined to one plane. If this be either of the principal planes of the rectangular rod, the more slowly or the more quickly travelling kind of disturbance, as the case may be, will alone be excited in the latter; and if the plane of the original disturbance be any other, the components into which we must resolve it in order to find the disturbance excited in the rectangular rod will in general be of unequal intensity, their squares varying with the azimuth of the plane of the original disturbance in accordance with Malus's law. This illustrates the subdivision of a beam of polarized light incident on Iceland spar into two of unequal intensity polarized in rectangular planes, and their alternate disappearance at every quarter of a turn. We see with what perfect simplicity the theory of transverse vibrations falls in with the elementary facts of polarization discovered by Huygens, standing in marked contrast in this respect with the conjecture by which Huygens himself attempted to account for double refraction.

But can we go further, and account for, or discover, from theory the laws of double refraction and the accompanying polarization in different directions in doubly refracting crystals?

It is to Fresnel we owe the first theoretical deduction of the laws of double refraction in Iceland spar,
and the discovery of the beautiful laws of double refraction in biaxal crystals ; laws some of which had been
previously known from observation, while in other
respects theory served to correct what had been
supposed to be the result of observation, but which
more careful observation carried out in directions indicated by theory proved to have been incorrect.   The
generalization by which Fresnel passed from the laws
of double refraction in uniaxal to those in biaxal
crystals is one of the most splendid things which has
been done in science.   And yet the theory which
guided him to the discovery of these laws is not one
which is rigorous throughout, nor did Fresnel himself
profess that it was, though in some reproductions of
his theory contained in text-books it is presented as
if it were, to the detriment of the student.   Fresnel
was a man of singular sagacity, endowed apparently
with a mind of the inductive class, leading him often
to the discovery of truth from conflicting or imperfect
evidence.   We may even say it is fortunate that
Fresnel did not rigorously follow out to their conclusions the premises with which he started, for had he
done so he would have missed the discovery of the
elegant laws of double refraction.   In fact, Cauchy
and Neumann independently worked out the con-

clusions which rigorously follow from the state of
things assumed by Fresnel ; but with all the squeezing
which the arbitrary constants furnished by theory
admit of, they were not able to obtain Fresnel's laws
except as an approximation.   Had this been all, it is
possible that the more complicated laws expressed by
their formulae might have fitted observation as well as
the simpler and more elegant laws of Fresnel.   But
the theory is hampered by a third ray which cannot
be got rid of, and which leads to conclusions at
variance with observation.   While admiring therefore
the geometrical part of Fresnel's theory, we must
reject the mechanical conditions which he supposed to
exist in crystals, as not being in conformity with
nature.

Fresnel's laws have been obtained in an extremely
elegant manner, as the result of a rigorous theory, by
Green, who starting with an assumed mechanical state
inclusive of but more general than that assumed by
Fresnel, and subsequently limiting the generality by
a single condition which the phenomena of light
would naturally lead us to introduce, arrived directly
at Fresnel's laws.   It was necessary however to suppose the vibrations of polarized light to be *in* the
plane of polarization.   In the same paper he showed
however that by starting with a mechanical state still
more general, limiting it as before, and introducing

two simple linear relations between the arbitrary constants remaining, Fresnel's laws were again obtained, but this time by supposing that in polarized light the vibrations are *perpendicular to* the plane of polarization. Almost simultaneously with Green MacCullagh obtained equations of motion of the ether in a crystal identical with those of Green in his first theory, though by a method which does not seem quite so satisfactory as that of Green. They led of course to the same laws, and required the same supposition as to the direction of the vibrations in polarized light. Lamé has given a theory substantially agreeing with Green's first theory; and last but not least we have the electro-magnetic theory of Maxwell, which without any straining or assumption of relations between constants leads directly to Fresnel's laws.

It may seem strange that we should arrive at the very same laws by such different theories; first by one which is not a rigorous theory at all, and then by others which are rigorous, but which differ among themselves, even to such an extent that in one the vibrations in polarized light must be assumed to be in, in another perpendicular to the plane of polarization. In explanation of this it is to be observed, first, that all these theories alike involve the idea of transverse vibrations, and secondly that Fresnel's laws are really the simplest which can in any way suit the phenomena. Fresnel's laws are embraced in an elegant construction applied to an ellipsoid; and just as an ellipsoid is the simplest generalization of a sphere when we pass from what is alike in all directions to what varies from one direction to another, so Fresnel's laws are really the simplest that the nature of the phenomenon, viewed in the light of the theory of transverse vibrations, admits of. It is not therefore so wonderful as at first sight might appear that the same laws should be arrived at from theories so different; and while the deduction of these laws is a strong confirmation of the truth of the theory of transverse vibrations which is common to all the methods, it is not itself alone to be taken as establishing the truth of the supposition as to the mechanical state of things in crystals which has been made in the deduction of the laws.

I have mentioned Malus's discovery of the polarization of light by reflection, and the question may naturally arise, is this reconcileable with the theory of transverse vibrations? To show that it is, I need only refer to Fresnel's deduction of the intensity of the light reflected from an isotropic transparent medium, according as the vibrations are in or perpendicular to the plane of incidence. These formulae were obtained by Fresnel with his wonted sagacity from a process only partially complete, since we must either allow that a part only of the necessary equations of condition at the common boundary of the media are satisfied, or else that the mechanical state for which the conditions employed give the complete solution remains to be defined.

Fresnel's deduction of his laws for the intensity of reflected light was made on the supposition that in polarized light the vibrations are perpendicular to the plane of polarization. A slight difference of hypothesis as to the state of things would lead, by reasoning very similar to that of Fresnel, to the very same two formulae only with the directions of vibration to which they respectively apply interchanged. On the present supposition therefore the vibrations in polarized light must be supposed to be in the plane of polarization.

The polarization of light through the unequal absorption belonging to the two rectangularly polarized pencils within certain coloured doubly refracting crystals, such as tourmaline, readily falls in with the theory of transverse vibrations. For there can be no doubt that absorption in general consists in the expenditure of the incident ethereal vibrations in producing molecular agitation; and it is easily understood that the molecules may be more easily agitated by an ethereal vibration in one direction than in another.

The aim which I have proposed to myself in this first course of the Burnett lectures has been to lay before you, as impartially as I could, a summary of the evidence on which we accept the answer to the question, What is Light? given when we say, Light consists of undulations in a medium, called ether, pervading the interplanetary and interstellar spaces, and existing also within bodies formed of ponderable matter. Difference of refrangibility, with the accompanying difference of colour, depends upon a difference in the frequency of these undulations. The direction of vibration of the particles of the ether is transverse to the direction of propagation of the light, and accordingly (at any rate in the case of vacuum or an isotropic medium) the vibrations take place in the fronts of the waves, but in common light are not otherwise restricted, while in polarized light they are rectilinear, taking place in a direction which is symmetrical with respect to the plane of polarization.

Naturally the full force of the evidence can be felt only by those who have well studied the subject. I hope however that I may have succeeded in showing, even to those who previously may have paid little attention to the matter, that at least there are powerful arguments in favour of the accepted answer. The inductive arrangement, which seemed best fitted for the object I had in view, naturally led to a treatment in good measure historical, but I have not, I hope, neglected recent researches. I have endeavoured to discriminate, and to lead you to discriminate, between what is well established and what is still speculative, and have confined myself almost entirely to the former.

Should I be permitted to deliver courses of lectures again in the two following years, it is my intention, in accordance with a scheme communicated in outline to and approved by the Burnett Trustees, to devote next year's course to researches in which light has been used as a means of investigation, while the third year's course would be assigned to light considered in relation to its beneficial effects. While the different objects to be held in view in these lectures are more or less blended together, the second course would more especially relate to recent researches, while the third would naturally harmonize with the original intentions of the Founder of the Trust.

END OF FIRST COURSE.

CAMBRIDGE: PRINTED BY C. J. CLAY, M.A. & SON, AT THE UNIVERSITY PRESS.

www.ingramcontent.com/pod-product-compliance
Lightning Source LLC
Chambersburg PA
CBHW021816190326

41518CB00007B/615